William McCartney

Olfaction and Odours

An osphrésiological essay

Springer-Verlag Berlin · Heidelberg · New York 1968

William McCartney, Ph. D., F.R.I.C., A.H.W.C.,
89 Arundel Road, Kingston-upon-Thames/Great Britain

ISBN-13: 978-3-642-87701-8 e-ISBN-13: 978-3-642-87699-8
DOI: 10.1007/978-3-642-87699-8

Title No. 1504

Foreword

If we possessed a true, well established and undoubted theory concerning the function of the nerves in the human body, the principles of smell would be much easier to understand. So far, we do not know for certain whether the action of nerves takes place by means of a subtle fluid, which is instantaneously conveyed from the beginning of the nerves to their ends, or whether, as some think, some sort of vibratory movements is produced in them, or whether the stimulation should be considered as the only sufficient cause, to say nothing of the new opinion concerning the electrical force.

Linnaeus

Although much has been learned since the time of Linnaeus, it is still true that we do not yet possess "a true, well established and undoubted" theory of olfaction. The literature of the subject, however, has long been rather voluminous — the frequent assertions to the contrary are simply untrue — and is now growing rapidly. Moreover, quite a number of symposia on olfaction (or olfaction and taste) have been held during recent years. For example, in the U.S.A., the New York Academy of Sciences, in collaboration with other organizations, has held two conferences *(Basic Odor Research Correlation,* 1953 and *Recent Advances in Odor: Theory, Measurement and Control,* 1963) and the American Society for Testing Materials held a similar conference, entitled *Symposium on Odor* in 1954; in Italy, at least one conference has been held by the Ligurian and Tuscan sections of the Italian Society of Laryngology, Rhinology and Otology; in Switzerland, the Society of Chemical Industry held its discussion of *Molecular Structure and Organoleptic Quality* in 1957; in Sweden, the first Wenner-Gren Symposium [1], held in 1962, was devoted to *Olfaction and Taste;* and in France and Italy since 1962 the excellent "Mediterranean" symposia on the sense of smell have taken place annually. [2] Nevertheless, no truly comprehensive modern monograph on olfaction exists although,

[1] The second symposium took place in Tokyo in 1965. And in 1966 two conferences on olfaction were held in the U.S.A. and one in Istanbul (Gordon Research Conference on Chemistry and Physiology of Odor and Flavor; Symposium on Odours, American Oil Chemists Association; NATO Advanced Studies Institute on Olfaction Theories and Odour Measurement).

[2] In 1935, a Smell Society was established in London but the second world war brought it to an untimely end. It may be added as an interesting pointer, that a large U.S. oil firm employs a "co-ordinating odor moderator" and that at least one reference to the profession of "odor engineer" has appeared in the literature.

as will be pointed out later, there are a few excellent works on the sense of smell in man or on the human senses generally and some valuable bibliographies have been compiled.

Since established facts, in the field of olfaction, are not numerous there has quite rightly always been considerable (although not infrequently unjustifiable) speculation. If most of this has been unfruitful, however, one reason is perhaps that it has not been well grounded on results previously attained. All too frequently, writers and experimenters have simply neglected to study the work of their predecessors or have studied it inadequately and at second hand using abstracts and reviews *in place of* instead of *as guides to* original literature. Moreover, despite the lip service paid to the international or cosmopolitan nature of science, many investigators and authors still rely for too much on works in their own language while seriously neglecting those in other languages.

Olfaction and Odours also is not comprehensive. It deals with birds, dogs, fish and man and its range of topics is otherwise selective. But perhaps it will arouse new interest in some much neglected aspects of science and, as Régismanset so charmingly suggested more than fifty years ago, it may stimulate someone more courageous, more acute and more learned also than the author to imitate him.

They haven't got no noses,
The fallen sons of Eve;
Even the smell of roses
Is not what they supposes;
But more than mind discloses
And more than men believe.

The brilliant smell of water,
The brave smell of a stone,
The smell of dew and thunder,
The old bones buried under,
Are things in which they blunder
And err, if left alone.

The wind from winter forests,
The scent of scentless flowers,
The breath of brides' adorning,
The smell of snare and warning,
The smell of Sunday morning,
God gave to us for ours.

And Quoodle here discloses
All things that Quoodle can,
They haven't got no noses,
They haven't got no noses,
And goodness only knowses
The Noselessness of Man.

G. K. Chesterton

«Deux hommes agés se rencontrent. Ils furent camarades de classe, mais la vie les a ensuite séparés. L'un d'eux a eu une carrière assez brillante pour que le Français moyen connaisse son nom. «Et toi, qu'as tu fait?» demande-t-il à son ancien condisciple. C'est une réponse simple: «J'ai passé ma vie à explorer le monde des odeurs et à créer quelques parfums.» Et il poursuit: «Tu fait la moue, comme s'il agissait de choses futiles. Tu ne peut savoir combien ce monde est vaste et l'attrait qu'il a pour moi. Chaque jour, je découvre de nouveau, et chaque découverte m'apporte une satisfaction de l'esprit. Dans la variété des sensations olfactives qui sont les miennes, dans leur qualité, je vois une preuve de la richesse, de la beauté de ce monde extérieur dont nous prenons conscience, nous, les parfumeurs, avant tout par l'odorat.»

E.-P. Meunier
(A true story.)

Acknowledgment

Grateful acknowledgment is made to the editors of the *International Perfumer* and *Soap, Perfumery & Cosmetics* for permission to incorporate in the present work material originally published in these periodicals.

Contents

Introduction

> This famous town of Mansoul had five gates in at which to come, out at which to go; and these were made likewise answerable to the walls, to wit, impregnable, and such as could never be opened nor forced but by the will and leave of those within. The names of the gates were these: Ear-gate, Eye-gate, Mouth-gate, Nose-gate, and Feel-gate. *Bunyan*

> Nous crûmes devoir commencer par l'odorat, parce que c'est de tous les sens celui qui paroit contribuer le moins aux connoissances de l'esprit humain. *Condillac*

How long a period of time has elapsed since the primate stock substituted sight for smell as the dominant sense we do not know, but it is certainly a very long time. Some investigators believe that this substitution occurred at about the same time as our ancestors took to life in the trees. In any case, it seems certain that, with the advance of time and civilization, the human sense of smell has degenerated. Possibly this is, to some extent at least, an advantage[1]. There is evidence, on the other hand, that the extent of the degeneration may actually be much less than is frequently supposed and it seems highly probable, as Parker has pointed out, that "the high development of this sense in our ancestry has left such a profound impression on the organization of our central nervous apparatus that we are often surprised by the power of our olfactory associations". And when we read accounts such as that given by Helen Keller in her chapter on *Smell, the Fallen Angel* we perceive, perhaps dimly, the immense value of this sense even to those whose other senses are unimpaired.

However that may be, it is undeniable that, to many people in scientific circles as well as outside them, the sense of smell is indeed a 'lower sense'. Very much time and labour have been expended, for example, in attempting to trace relationships between chemical constitution and physiological action. And although these attempts have so far yielded only meagre and sometimes rather disappointing results, many workers are nevertheless continuing the search. It is surprising, therefore, that so little attention has been directed by scientists (poets and men of letters have perhaps given it more attention)

[1] "The sensations most heavily charged with emotional content are sensations of smell. Man's sense of smell is relatively poor and this apparent handicap has proved to be an actual advantage to him. Instead of running round like a dog, sniffing at lamp-posts and becoming deeply agitated by what he smells on them, man is able to stand away from the world and use his eyes and his wits relatively unmoved." A. Huxley.

to the sense of smell. No doubt some have felt that the subject is one for mirth rather than for serious investigation. These will be surprised to learn that, not so very long ago, Dumas made a scientific study of the 'odour of sanctity' said to be possessed by many mediaeval saints and even attributed to the substance (or substances) responsible for it the formula $H_6H_{12}O_2$. Some of the saints, he suggested, may have suffered from diabetic acetonuria and consequently may have smelled of butyl acetate, propyl propionate, ethyl butyrate, or some similar compound [2]. Others may have been repelled by obstacles in the path of research into this sense [3]. Was not Professor Nencki, during the period of his studies on skatole, asked to leave the reading room of the Museum Society of Bern? Did not passengers in the omnibus in Burton edge away from Peter Griess when he was investigating certain putrefaction products? ("Ladies, it is not what you think it is!", he remarked.) And, only a few years ago, were not some workers from Ruzicka's laboratory in Zürich expelled from public transport vehicles because of the disagreeable steroid odour which clung to them? McKenzie, in his book *Aromatics and the Soul,* sang of "scents, perfumes, odours, whiffs and niffs; of aromas, bouquets and fragrances; and also ... of efffluvia, reeks, foetors, stenches, and — stinks". But apparently he did not himself work with the latter! Difficulties alone, however, can scarcely account for the general neglect of this subject which, let it be noted, has been studied, not very extensively in every case, by men of the eminence of Adrian, Barcroft, Emil Fischer, Ramsay and Ruzicka ("the little rose") to mention but a few. And perhaps those who mock may be silenced by a request to explain even one of the profound mysteries and baffling problems of olfaction. Surely it is an astonishing (or should one say alarming?) thing [4] that one gramme of an odorous substance could suffice to provide an olfactory impression for the whole population of the world, and that the olfactory powers of a moth should extend almost to sub-atomic limits [5]. Is there anyone who can fail to be amazed at the almost incredible ability

[2] Repond, however, has described a case which showed that the 'odour of sanctity' was possibly much more complex than this.

[3] "Il est si facile à nos grands savants de laisser tout simplement tomber une question quand elle les embarrasse. Mais, au fait, pourquoi l'ont-ils laissé tomber???" R. Boucher.

[4] Bei manchen Duftkörpern würde ein Gramm ausreichen, um sämtlichen (1,6 Milliarden) Bewohnern unseres Erdballes eine eben merkliche Geruchsempfindung zu verschaffen. Eine scharfe Nase bemerkt die Anwesenheit eines ausgiebigen Riechstoffes 100 000mal früher, als sie sich spektralanalytisch und chemisch nachweisen läßt. Daß der Geruchssinn unser *empfindlichster* Sinn ist, steht danach ganz außer Frage." H. Henning.

[5] The honey bee, on the other hand, seems to have about the same olfactory acuity as has man although, for substances of biological importance to bees (*e. g.*, beeswax) it is much superior to him. Conversely, to the odour of oil of rosemary, man is apparently five times as sensitive as is the bee.

of some animals to detect friends or enemies, often at great distances, apparently by means of the sense of smell almost alone? And if, as we seem obliged to suppose, these powers depend in part on the transport of material particles, what kind of process is it that enables the animal to detect such infinitesimal amounts of material? Further, what kind of reaction, chemical, physical or physiological occurs at the olfactory mucous membrane so that the nervous impulse is initiated? To these questions and to many others like them, no satisfactory answer can be given and it seems that few serious attempts have been made to provide one. Yet it is quite possible that a determined attack on these problems, even if it were only partially successful, might have valuable results since, not inconceivably, it might throw light on fundamental questions of, for example, molecular and atomic structure. Again, a solution of the physiological riddles involved could scarcely fail to be of immense value to physiological and medical science.

It is clear then, that the study of the mysteries of odour and of the sense of smell could easily be justified if justification were required. More probably, however, the study will be made, if indeed it is made at all, to satisfy the curiosity of those who will not be troubled by the mockery of the onlookers. "La philosophie des parfums", exclaims Régismanset in his delightful little book bearing the same title, "Oh sans doute, vais-je être taxé de folie! Hé quoi, cette rose épanouie, cette grappe gracile de jasmin auraient quelques rapports avec la sagesse? Et puis, pourquoi toujours chercher les fins des choses? Leur spectacle ne nous suffit-il pas, qu'il faille encore étudier leur but et leurs causes?"

The philosophy of perfumes has doubtless been neglected. Perfumery certainly has not and it is not surprising that some of the best work on olfaction is done by perfumers, although, if Janistyn is to be believed, they rarely create really new perfumes — only variations of existing ones. Workers in fields other than that of perfumery also study practical aspects of olfaction. Thus, food manufacturers, packaging experts and salesmen learn how to prevent wrappings from acquiring unpleasant odours and how to increase sales of clothing and of other goods by giving them attractive odours (which may be only sub-consciously perceived). Chemists find how to produce evil-smelling but none the less valuable warning agents and how to deal with stinks which would otherwise travel long distances from their places of origin. Botanists show that plants yielding scentless flowers can be caused to produce scented ones by appropriate breeding or by treating seeds or roots with odorous material [6]. Entomologists discover how to control insect pests by studying their olfactory powers and producing suitable attractants and repellants. And military men utilize their sense of smell in detecting war gases. In short, if the search for "goals and causes" goes slowly, work on the practical applications proceeds apace.

[6] Heyl: U.S. Patent 2,168,523. 1939.

Historical

In einem Aufsatz der „Deutschen Revue" habe ich die Behauptung aufgestellt, die Seele entdeckt zu haben, indem ich einen ganz bestimmten chemischen Bestandteil des Körpers als Seele denunzierte, nämlich jenen Stoff, bzw. jene Stoffe, welche die völlige Spezifität des Ausdunstungsduftes und des Fleischgeschmacks bedingen. *G. Jaeger*

Perhaps the first treatise on odours is that of Theophrastus who wrote his *Concerning Odours* in the second or third century B.C. He maintains that odours "are due to mixture for anything which is uncompounded has no smell, just as it has no taste". Everything, plant, animal or inanimate object that has an odour has one peculiar to itself. Often we cannot perceive the odour since our sense of smell is inferior to that of all other animals and they can smell things which appear odourless to us. Theophrastus also points out the kinship between taste and smell. These senses, he says, facilitate the enjoyment of each other and he claims that two odours may combine to form one. He remarks on the fact that some substances retain their odours indefinitely or at least for very long periods and he discusses olfactory exhaustion, the adsorption of odorous substances, the classification of odours and many other interesting points.

It may be supposed that, in the following centuries, man's interest in olfaction did not wane although it seems that for long little or nothing was written on the subject. In any case, in the early years of the XVth century we find Thomas Norton of Bristoll, in his *Ordinall of Alchimy,* that

> *trusty-Booke* of *faithfull-Things*
> Instructing *Priests,* Advising *Kings*

devoting not a little space to smelling, odors, stinkes, sweete Savours, and the like. He tells us that

> *Odor* is a smokish vapour resolved with heate,
> Out of substance, by an invisible sweate;

and that

> *Stinch* is a Vapour, a resolved fumositie
> Of things which of Evill Complexions be.

Montaigne, nearly 200 years later, wrote an essay *Of Smells* in which he informs us that the cooks in the service of the King of Tunis "stuffed his meats with odoriferous drugs with such extravagance that the cost of dressing a peacock and two pheasants after their fashion amounted to a hundred ducats; and when they were carved up, not only the banqueting hall, but all the rooms in his palace, and even the neighbouring houses, were filled with very sweet fumes, which did not vanish for some time after". Obviously, such odoriferous extravagance could not escape the attention of Montaigne's contemporaries. We know from the work of Kenneth that one of these, Camerarius by name, presented to the University of Marburg in

1587 a thesis entitled *Themata physica de odorum natura et affectionibus* in which be discussed the mechanism of olfaction, the possibility of smelling in water, the classification of odours (heavy, acute, fragrant), the relationship between taste and smell, and the effect of solar heat on odours.

In England we find, not unexpectedly, that Robert Boyle (1627—1691) directed his attention to the subject. In his *Experiments and Observations about the Mechanical Production of Odours,* for example, he relates the following. "An eminent professor of mathematics affirmed to me, that, chancing one day in the heat of summer, with another mathematician (who, I remember was present when this was told) to pass by a large dunghil, that was then in *Lincoln's-Inn Fields,* when they came to a certain distance from it, they were both of them surprised to meet with a very strong smell of musk (occasioned, probably, by a certain degree or a peculiar kind of putrefaction), which each was for a while shy of taking notice of, for fear his companion should have laughed at him for it; but, when they came much nearer the dunghil that pleasing smell was succeeded by a stink proper to such a heap of excrements." Choosing civet as an example, he proceeds to discuss the peculiar fact that a substance which has a pleasant odour when highly diluted may have an indifferent odour or a very repulsive stench when concentrated. And, like Montaigne, he mentions that the perspiration of some persons and of some animals has a fragrant odour. "An ingenious lady, to whom I am nearly related", he also tells us, "showed me an odd monkey, that had been presented her as a rarity by the then admiral of *England,* and told me, among other things, that she had observed in it, that, being sick, he would seek for spiders as his proper remedies, for some of which he then seemed to be looking, and thereby gave her occasion to tell me this; which, when he had eaten, the alteration it made in him, would sometimes fill the room with a musky scent: but he had not the good luck to light on any, whilst my visit lasted." Boyle was also familiar with the use of fixatives, which seem to heighten and improve the odours of other substances. He discusses the use of 'amber-grease' and points out that "just proportion and skilful mixture" are as important as choice of odorous substance in the compounding of noble perfumes. In another series of essays, those on *Strange Subtilty, Great Efficacy, Determinate Nature of Effluviums* he comments on the strange extensibility of some bodies, whilst their parts yet remain tangible, the multitude of visible corpuscles that may be afforded by a small portion of matter, the smallness of the pores at which the effluvia of some bodies will get in, the small decrement of bulk, or weight, that a body may suffer by parting with great store of effluvia, and the great quantity of space that may be filled, as to sense, by a small quantity of matter when rarefied or dispersed. He showed by experiment that ambergris and asafoetida could "part with good store of odoriferous steams" during periods of several days without losing measurable weight and he noted that his pair of Spanish perfumed gloves retained their odour for nearly thirty years. (It is said that Newton made a similar observation.) Amongst the philosopher-poets of Boyle's age was Sir John Davies who fluorished some decades earlier than the father of modern chemistry. Davies' poem, *Nosce Teipsum (On the Soul of Man and the Immortality Therof)* includes these lines:

Next in the nosthrils she doth use the smell:
As God the breath of life in them did give,
So makes He now this power in them to dwell,
To judge all ayres, whereby we breathe and live.

This sense is also mistresse of an Art,
Which to soft people sweete perfumes doth sell;
Though this deare Art doth little good impart,
Since they smell best, that doe of nothing smell.

And yet good scents doe purifie the braine,
Awake the fancie and the wits refine;
Hence old devotion, incense did ordaine
To make mens' spirits apt for thoughts divine.

The scientific writings of the XVIIIth century also include not a few dissertations on olfactory subjects. Albrecht von Haller, for example, reported that he had preserved, for more than forty years, 400 dissertations of forty pages each which a single grain of ambergris had perfumed. At the end of that period, the pages had lost nothing of their odour although, as the learned physiologist calculated, each square inch of the surface of the paper had been impregnated with only 1/2,691,064,000 of a grain of the substance. Towards the end of this century, too, Lorry published his *Observations sur les parties volatiles et odorantes des médicamens tirés des substances végétales et animales* and, about ten years later, in the years V and VI of the Republic (1797 and 1798) to be exact, we find Citizens Prévost and Fourcroy writing in the "*Annales de Chimie*" papers entitled *Sur les émanations des corps odorans* and *Mémoire sur l'esprit recteur de Boerrhaave, l'arome des chimistes français ou le principe de l'odeur des végétaux.* This mysterious *Esprit Recteur,* it may be remarked, was defined by Fourcroy as "être très-volatil, très-fugace, très-expansible, privé presque de pesanteur appréciable et complètement invisible, intactile, ou seulement insensible à d'autres organes que la membrane olfactive, l'*esprit recteur* paroissoit se ranger de lui-même au nombre des fluides élastiques". But it was in the early part of the XIXth century that, in France at least, workers in the domain of olfaction became especially active as witness, for example, the works of Godineau (1809), Kirwan (1808), and Geoffroy Saint Hilaire (1825). It was in this period indeed that there appeared one of the few full length discussions of olfaction ever published. This is the exhaustive treatise by Hippolyte Cloquet, M. D. entitled *Osphrésiologie ou Traité des Odeurs du Sens et des Organes de l'Olfaction* issued at Paris in 1821 [1] (Second edition. The first edition, which appeared in 1815, was, as the author says, only a simple dissertation. But a review in the first volume of the *Journal de Pharmacie* (1815) asserts that it was "not a simple thesis but a complete treatise."). Cloquet, perhaps, was not a physiologist of the first rank. But how many physiologists, or for that matter, how many scientific workers in any field today could truthfully make claims such as those made by this author in the preface to his book. "Aussi je n'ai considéré l'accueil favorable de mes confrères que comme un encouragement pour faire mieux; j'ai approfondi, j'ai medité mon sujet, j'ai fait en sorte de changer une simple dissertation en une monographie complète. J'ai visité les bibliotheques, j'ai parcouru les hôpitaux, j'ai vérifié sur les cadavres tous les détails anatomiques, j'ai rassemblé et coordonné tous les faits connus jusqu'a moi, je n'ai fait qu'en déduire les consequences qui en découlent immédiatement et j'ai repoussé au loin toute hypothèse hasardée, toute théorie non encore prouvée ... je ne dis point de n'en commettre aucune erreur, mais au moins de n'en pas commettre de dangereuse." Truly, an indefatigable and scrupulous worker!

In the half-century following the publication of *Osphrésiologie* little or nothing on olfaction was published. The very interesting experiments of Emil Fischer and his collaborator Franz Penzoldt were carried out about the year 1886, it is true, a few years after Gustav Jaeger had 'discovered the soul'. But it was in 1892 that perhaps the first study of the sense of smell by modern methods was made. The author of this study, *Fiziologiya nervi olfactorii istoricheskiya i eksperimentalniya izsliedovaniya,* was Nikolai Aleksandrovich Saveliev subsequently professor in the University of Yuriev, who, in the following year, published a similar (or possibly identical) work entitled *Fiziologiya obonyaniya istoricheskiya i eksperimentalniya izsliedovaniya* (Physiology of smell considered historically and experimentally) in the Reports of the physiological laboratory of the University of Moscow. These seem to be outstanding works but unfortunately they appear to be almost unknown outside Russia.

About this time, Zwaardemaker senior, later professor of physiology in the University of Utrecht, began his long-continued investigations of odours and the sense of smell. His famous book, *Die Physiologie des Geruchs* (1895), was apparently written in the Dutch language but was translated into German before publication. (It is worthy of note that, as late as 1951, Foster, commenting on the need to study energy source, intensity, quality, classification, affectivity and other matters, pointed out that little or no

[1] A German translation appeared in 1824.

advance in olfactory research had taken place since Zwaardemaker began his work about 1890.) Much later (1925) a French version of the book, unfortunately not showing much advance on the German work, appeared in Paris. After this, the experimental study of olfaction was not left, as it had so far been, almost entirely to chemists and physiologists. Psychologists, particularly in Germany and the U.S.A., began to turn their attention to the subject. Henning's work, *Der Geruch* is perhaps the best evidence of this. Despite the fact that he was told "there is nothing to write about smell", he produced a monumental treatise (first edition, 1916; second edition, 1924) claiming to be a *Handbuch für die Gebiete der Psychologie, Physiologie, Zoologie, Botanik, Chemie, Physik, Neurologie, Ethnologie, Sprachwissenschaft, Literatur, Ästhetik und Kulturgeschichte!* This indispensible book, still perhaps the most important of the psychological works, is to some extent marred by many unsupported claims and material of doubtful relevance. Zwaardemaker, with much justification, subjected it to severe criticism. It offers, he said, a series of observations which, for the trained physicist, are superfluous and, for the psychologist, have the disadvantage that the purely empirical matters with which alone they deal are buried under a mountain of in-applicable detail. Henning, he continued, was so much of a psychologist that he did not appreciate the value of physiological experiment. In the experimental field, where his technique was deficient, he heaped "Ungeheuerlichkeit auf Ungeheuerlichkeit".

Other modern works on olfaction are not numerous and the more important are quickly enumerated. Some of them seem to be seldom consulted.

First comes one of the classics, the *Essai d'Olfactique Physiologique* (Bruxelles: 1919) by A. Heyninx. He advanced a vibratory theory claiming that odorous molecules are those having specific ultra-violet vibrations of wave-length between 0.35 μ and 0.20 μ, this being the human range. In this range fall the "seven fundamental odours", namely, sharp, rotten, foetid, burnt, spicy, vanilla and ethereal. The molecules become adsorbed on the olfactory mucous membrane and initiate vibrations in it, these vibrations being transmitted to the olfactory cells. Olfactory pigment probably acts by olfactory resonance to intensify the vibrations of the odorous molecules.

A very different but also very good general work on the anatomy and physiology of the olfactory system is G. H. Parker's *Smell, Taste and Allied Senses in the Vertebrates* (Philadelphia and London: 1922). This was followed by E. von Skramlik's *Handbuch der Physiologie der niederen Sinne. I. Die Physiologie des Geruchs- und Geschmackssinnes* (Leipzig: 1926), a scholarly work in which nearly 350 pages are devoted to smell.

A very different approach to the subject was that of the late Jean Nogué whose posthumous work *Esquisse d'un Système des Qualités Sensibles,* published in Paris in 1943, includes fifty pages on olfaction.

Nogué's standpoint is perhaps best revealed by remarking that his book appeared in the Bibliothèque de Philosophie Contemporaine.

Different again is R. W. Moncrieff's *The Chemical Senses* (London: 1944. Second edition, 1951. Third edition, 1967.). This is certainly not a philosophical work.

The first and perhaps the only full length American scientific work on olfaction is C. P. McCord and W. N. Witheridge's *Odors: Physiology and Control,* a valuable and comprehensive publication. These two authors claim that they "advance a new theory of odor stimulation". However, the details of this electro-chemical theory are not at all clearly set out and are even said to be based, frankly, on speculation. The suggestion made seems to be that odour depends on "modification of bonding angles between atoms", this modification occurring when the atomic shape is distorted on dissolution in the olfactory mucosa. Properties which confer odourlessness are given but not those which confer odorous power.

McCord and Witheridge's book was published in New York in 1949. In the following year, the Academy of Science of the U.S.S.R. issued Bronstein's study of taste and smell entitled *Vkus i Obonianiye.* This is possibly the best and most comprehensive, if not the only, modern Russian scientific book on these two senses. It contains a useful bibliography of Russian and other publications.

One year later, again, Jellinek's *Die psychologischen Grundlagen der Parfumerie* appeared in Heidelberg. Much of this book is devoted to the author's rather doubtful theories concerning the 'erogenic' and 'anti-erogenic' action of perfumes.

A second work on the chemical senses, Kalmus and Hubbard's *Chemical Senses in Health and Disease* was published in London in 1960. The similarity of the title to that of Moncrieff's book does not extend to the contents which, moreover, are rather disappointing because of their brevity and their omissions. Nevertheless, within its limits, it is a valuable and up-to-date work which poses such questions as: why is the scientific exploration of taste and smell in such a deplorable state and why are the cook and the perfumer still artists and not scientists?

Finally, we have a useful book from Italy, namely, Fortunato and Niccolini's *L'Olfatto, Cognizioni Antiche e Recenti* (Roma: 1958), a lengthy and up-to-date work for the anatomist, biologist, embryologist, neurologist and physiologist and *Olfaction and Taste* which constitutes the proceedings of the first international symposium held at the Wenner-Gren Centre, Stockholm, in September 1962[2].

All these are chiefly, if not entirely, works of pure or applied science. But there are also several works which fall into the category of belles-lettres.

[2] R. H. Wright's excellent work, *The Science of Smell* (London: 1964) should now be added here and also *Kyukaku no Rinsho* by Masao Ichihara (Tokyo: 1963).

These include Régismanset's *Philosophie des Parfums* (Paris: 1907), Sansot's *Essai sur les Parfums* (Paris and Nice: n. d.), McKenzie's "light omelette of scientific literature" *Aromatics and the Soul* (London: 1923), Monéry's *L'Ame des Parfums* (Paris: 1924), Le Gallienne's beautiful booklet *The Romance of Perfume* (New York and Paris: 1928), Estève's *Parfums et Belles-Lettres* (Paris: 1939), Blaizot's *Parfums et Parfumeurs* (Paris: 1946) and Faber's *Zärtliches Brevier der Düfte* (Wiesbaden: n. d.). In Monéry's finely produced book, an "essai de psychologie olfactive", he supports the vibrational theory and, amongst many other speculations, he surmises that Milton's frequent use of olfactory imagery *(e. g.,* in *Paradise Lost)* may have resulted from his blindness.

Finally, there are a few popular or semi-popular works not all of which are, scientifically, very sound. They include: Sagarin's *The Science and Art of Perfumery* (New York: 1945), Bienfang's *The Subtle Sense* (Norman, Oklahoma: 1946), Millen's *Your Nose Knows* (Los Angeles: 1960) and Bedichek's *The Sense of Smell* (New York: 1960).

Perhaps it should be added that there are very few modern works devoted to olfaction in animals. Several are mentioned in the following pages.

In 1947, R. Boucher dolefully maintained, in a very interesting paper, that osphrésiology has become a "science of the past" and remarked that the word is not now to be found in the dictionary. And he added that difficult osphrésiological problems are still encountered by medical men and others. They are. But it is scarcely the whole truth that osmics (a modern name for osphrésiology) is a science no longer studied.

The results of the studies, some of which are discussed in the chapters which follow, are to be found, of course, chiefly in the great mass of material published in the medical, scientific and technical periodicals of all the advanced countries of the world. Here, it seems worth while to draw attention to a few general papers which, for one reason or another, are of special interest.

The first of these is Gustav Jaeger's "On the significance of taste and smell" which appeared in 1876. Man, he said, has two extraordinarily fine observing instruments, namely, his senses of taste and smell. With these he can distinguish, even when blindfolded, various animal species by smell, that is, by the species odour derived from perspiration. Each, race, variety and even individual has its specific odour and animals select mates, recognize offspring, other individual animals and animal species by scent. Scent is found particularly in excretions and it depends, in part, on the kind of food eaten. However, different species retain their species odour even when they consume the same kind of food for years. Jaeger observed a great many species in a zoo and he maintained that specific scents are inherited. Despite his peculiarities, his occult opinions and his strange beliefs about clothing, he was an acute observer whose views should by no means be disregarded. An example of his odd outlook is to be found in his British Patent No. 6001 of 1884. This patent, which deals with "Improvements in the preparation of food and other articles for medicinal purposes", describes the preparation by "neutral analysis" of a scent-extract from the hair of healthy women. The hair is cut into small pieces and ground, first with lactose, then with water and finally with alcohol. Homoeopathic dilutions of this extract are said to be "energizing and animating" and advantageous to health. They are to be added to food or perfumes.

One of the few early psychological treatises on olfaction appeared in 1894. The author, C. M. Giessler, had been told, as Henning was, that there was nothing to write on olfaction. One of the puzzling facts which he mentions is that terror is produced by the odours of lions and tigers in dogs and horses which have never seen these wild animals or previously encountered the odours.

The next paper to be noted was published more than twenty years later. The author, E. L. Backman of Uppsala, applying Overton's distribution law, came to conclusion that, in homologous and isomeric series of chemical compounds, intensity of odour increases as solubility in water decreases and solubility in lipoid matter increases. When the concentrations of odorant outside and inside the olfactory cell are equal, odourlessness results. He claimed, also, that some inorganic and organic substances, usually considered to be odourless, nevertheless have odours perceptible when the compounds are dissolved in physiological salt solution and passed into the olfactory cleft. Other subjects concerning which he made useful observations were olfactory adapation and compensation (or neutralization). Backman's work seems to have been overlooked possibly because it appeared in Sweden while the first world war was in progress.

Börnstein, who wrote more than ten years later than Backman, devoted his attention chiefly to the psychology of olfaction and to the place of the sense of smell amongst the human senses. His interesting and rather philosophical paper does not lend itself to condensation. He maintains that those who claim for smell a high place in the hierarchy of the senses are mistaken. It *is* a lower sense.

Achelis, whose thesis on „Geruchsstudien" appeared soon after Börnstein's paper, also dealt to a great extent with psychology. His views are of special interest in connection with those of Herrmann which will be discussed later and he expressed a large measure of agreement with Herrmann. Today, the work of Börnstein and Achelis, like that of Backman, seems to have been forgotten.

Also forgotten, apparently, is A. F. Baker's lecture on *„The significance of smell: some speculations on the phenomenology of olfaction"* which was given sixteen years after the war at the Royal College of Science in London.

The great value of this outstanding contribution is perhaps best illustrated by quoting the author's conclusions:

1. Smell, in contradistinction to Sight, is refractory to geometrical analysis and spatial location.

2. This by no means betokens a diminished sensibility.

3. The relations of smell and sight to organic process are antipodal; whereas the latter has regard to actualities extended in space, the former has regard to potentialities enduring in Time.

4. The relation of smell to the enduring potentialities embraces the extreme moments of Duration.

5. Smell is related to the Past through Memory; smell evokes memory. It is related to the Future through its function as an aphrodisiac; smell excites sexuality.

6. In the latter case, the subjective relation appears objectively through the relation of the gonads to the smell-producing organs.

7. The existence of smell in the present is fugitive; its *intensity* is balanced by the readiness with which it is *fatigued.*

8. Smell is a chemical sense; it intimates the relations of substances to the ascending and descending phases of organic metamorphosis.

9. This relation is illustrated by the reciprocity of Stinks and Perfumes.

10. Sight, which is superficial, is concerned with radiations whilst Smell, which is profound, is concerned with particles.

11. Smell, by its evocation of potentialities, institutes bonds between individuals, and so lures associations into spatial actuality.

12. These individuals may be of the same or different species; the associations, that is, genetically homogeneous respectively. The former have as archetype Sexual, the latter Symbiotic relations.

13. These relations are illustrated by the ambivalent rôle of smell in the inauguration, simultaneously, of the intersexual bond between Lepidoptera as homogeneous and the inter-floral bond as heterogeneous associations. Thus moths and butterflies tend, other things being equal, to prefer those flowers whose perfume resembles their own sexual smell.

14. Smell and Taste have their primordial matrix in a Chemotropic Sense in which both are blended, neither distinct.

15. So regarded, the evocatory rôle attributed to smell is perhaps susceptible of generalisation; the validity of this generalisation could be tested by experiment.

The conclusions reached, a year later by Nogué, author of the philosophical work mentioned above, lack the precision of Baker's but provide a very fitting summary of his "*Essai d'une Description du Monde Olfactif*".

The approach of Foster, whose Cornell University thesis is entitled "*A quantitative and qualitative analysis of odors*", differed greatly from those of Baker and Nogué. Using a large olfactorium, he conducted a great number of experiments in which about 350 compounds were examined chiefly from the psychological point of view. The thesis was submitted in 1949.

In the year 1953 several very useful reviews of olfaction were published. From Belgium came Hennebert's communication about one hundred pages long, which gives a good summary of modern knowledge and from France, reviews by Gerebtzoff and Le Magnen in the *Journal de Physiologie* which, though less comprehensive perhaps, are certainly not less informative. Hennebert's lengthy bibliography is unfortunately marred by a large number of typographical and other errors.

It was in the year 1953 also that the New York conference on "*Basic Odor Research Correlation*" was held. The twenty-six papers then submitted cannot be summarized here and it may suffice to mention one only, that of the eminent biochemist J. B. Sumner who dealt with a possible relationship between enzymes and the perception of odours and, with considerable hesitation, advanced a tentative theory which rather closely resembles the osmoceptor hypothesis of Ruzicka. Sumner also commented on the energy required for olfactory stimulation, on the role of the biochemist in olfactory research and on radiation theories. Such theories, he felt, are unconvincing if not actually preposterous.

The second American conference, a "Symposium on Odor", held considerably less than two years later, was much narrower in scope, its main theme being the prevention and control of obnoxious odours. Of the six papers submitted to this gathering, only one need be mentioned here. This is the very valuable communication by Deininger and McKinley entitled "*The design, construction and use of an odor test room*", a room which does not differ essentially from Foster's olfactorium but, as would be expected, is much superior to the cubicle described by Emil Fischer and Franz Penzoldt seventy years before.

Mullins, the next writer to be mentioned, published an outstanding contribution to olfactory research in the following year (1955). Working with highly purified *n*-butane as standard test substance for threshold measurements and with other simple chemical compounds (*e.g.*, alcohols, acids, esters) he came to the conclusion that there must be two or more types of receptors in the olfactory epithelium and that the odour of a chemical compound must be determined by the internal attractive forces in its molecules as well as by their size and shape. Odorous molecules possibly act by producing local disorder in the oriented molecular structure of the cell membrane.

The symposium on molecular structure and organoleptic quality, held in Geneva in 1957, was organized by the overseas section of the Society of Chemical Industry. Unlike the American conferences it gave due weight to the views of authorities from several countries without neglecting the work of the pioneers. Amongst the eight papers read at this symposium, those of Beets, Wright and Thompson are perhaps of greatest interest. In his discussion of the relationship between structure and odour in chemical compounds, Beets rendered very great service by emphasizing that "when the structure of an unknown organic compound is drawn with all substituents in their correct positions and with complete indications of configurational details, we have at the same time irrevocably and fully defined all the properties, chemical, physical and physiological, including the odour which the compound will have as soon as it has been synthesized". For his part, Wright gave an up-to-date account of his vibrational theory and Thompson, commenting on theories of smell, maintained not only that Wright's views but also those of Dyson, Beck and Miles, cannot withstand rather severe criticism. It is noteworthy that no less an authority than Guillot, in his review of the symposium, directly rejected Beets's affirmation. Certain steroids, he pointed out, are odourless for some normal people but not for others. Moreover, these compounds have quite different odours for different people. Hence, he argued, there is no direct relationship between molecular structure and odour. (It could be replied, however, that the variable factor here is not the property of the compounds but the biological condition of the perceiving subjects.)

Guillot has since elaborated his views on many occasions. He has recorded his experiences of the powerful psychological and physiological effects of odours even when they are unrecognized and, apparently, present in subliminal concentrations; he has praised Ehrensvärd's interfacial theory [3] and rejected that of Crocker and Henderson [4]; he has discussed specific parosmia, *e. g.*, to exaltolide and the dozen or so compounds which behave like it and he has supported the theory that optical asymmetry is a property not only of many odorants but also of the material of the nerves. When a substance exists in optically active forms, one of these may be two or three times as odorous as the other even when both are prepared from the same inactive form. In some cases, one form may be odorous and the other odourless. Contact between odorant and receptor is essential and if the degrees of adsorption of the two forms on the receptors differ, than, as with isoborneol for example, the *quality* of the odours will also differ. There must be at least a dozen differently specialized receptor cells. Some of the steroids which behave like exaltolide have at least three odours which depend on the age and sex of the percipient. (Those which have very unpleasant odours for some people were responsible for the expulsion of certain laboratory workers from public transport in Zürich.) Guillot has also reported a very curious phenomenon which he and some others have observed. It is

---

[3] See p. 162.

[4] See p. 116.

that olfactory acuity is often much greater or much less in the U.S.A. than in other parts of the world (Mexico, for example). The cause of this is not known but the factor concerned is not temperature or humidity.

Only two other publications, both very recent, will be mentioned here. They are Stuiver's Groningen thesis (1958) on the biophysics of the sense of smell which is valuable and up-to-date if somewhat too theoretical and Ottoson's very modern review (1963) of some aspects of the function of the olfactory system which, although rather deficient from the practical and psychological angles, gives, amongst other matters, a very helpful account of the most recent methods of investigation such as electro-physiological examination.

Dog's Life

Indem also der Schöpfer dem Hunde einen besonders feinen Geruch verlieh, gab er ihm die Möglichkeit, momentane und temporäre Vereinheitlichungen seines Wesens zu erlangen, welche ihn auf eine höhere Stufe im Tierreich erhoben und ihn geeignet machten, den Menschen zu verstehen, ihm zu dienen und ein treuer Lebensbegleiter desselben zu werden. *Giessler*

Vielleicht wird es dem Leser aufgefallen sein, das ich bisher von der Geruchsfähigkeit des Hundes gar nicht gesprochen. In der Tat sind hier die Experimente darum so schwierig, weil eben der Hund auf diesem Gebiet weit mehr versteht als wir Beobachter; das Unterscheidungsvermögen des Hundes für Gerüche ist ein so unendlich viel feineres als das des Menschen, daß wir dem Hunde in dieser Beziehung überhaupt nicht imstande sind zu folgen und jedenfalls kaum mit ihm experimentieren können, weil er eben offenbar Unterschiede macht, von denen wir uns nur schwer eine Vorstellung machen können; möglich wäre es ja vielleicht, daß ein williger und anstelliger Experimentator in dieser Beziehung allmählich vom Hunde etwas lernte, und durch Übung jene von dem Hunde gemachten Unterschiede auch erkennen lernte. *G. F. Nicolai*

The olfactory powers of some animals have naturally attracted considerable, if usually casual, attention and scientific studies of olfaction in various species have been conducted from time to time. Accurate observations of most mammals, however, seem to have been made rather infrequently. Lancum has recently recorded that the badger has an extremely acute sense of smell. It has also been reported (from the London Zoological Garden) that lions, leopards and other members of the cat family greatly enjoy the odours of lavender, rose, pinks and lilac, whilst Burton has remarked that the olfactory sense of tigers is very highly developed. On the other hand, it has been noted (in one case only, perhaps) that the lion may lie down with the hunter — within three feet of him, in fact — apparently without smelling him. Both these findings are somewhat surprising. Numerous field observations have been made by huntsman (they say, for example, that red deer can scent man *in still air* at more than a hundered metres and that the kangaroo can scent man and man scent skunk at a distance of four miles) and by those interested in the tracking of law-breakers. But it is remarkable that exact experiments have not often been conducted and that, indeed, purely scientific studies of the question seem to have been made rather seldom. More interest has probably been taken in the dog than in any other higher animal. In any case, few if any animals are so intimately associated with man as are dogs and this chapter will be devoted to them [1].

The olfactory power of the dog and also its mental ability have probably been over-estimated (not seldom ludicrously overestimated) much more frequently than they have been under-estimated. He is obviously a very intelligent animal (he probably understands the human world much better than man understands the canine) although his intelligence, it must never be forgotten, differs wholly or in large measure from human intelligence. Thus, the most skilful and highly trained police dog will perform astonishing tracking feats but he can have little or no conception of the human purposes which he serves. Often, however, this fact has been forgotten or even denied with the result that the life and liberty of suspects or accused persons (escaping negro slaves, for example) have been risked, on many occasions quite recklessly, on the basis of blood-hound 'testimony', some law courts having accepted the trustworthiness of canine evidence as 'common knowledge' and ignored the fact that a canine 'witness' cannot be cross examined. Other law courts (*e. g.*, in Germany at one time and today in the U.S.A.) have refused to accept any evidence at all based on the work of dogs, maintaining, perhaps, with Craig, that they have been completely or largely discredited as witnesses. Although this is almost certainly not the case, it is equally certain, as several German authors have stressed and as McWhorter has indicated in the *American Law Review* and Craig in the *Scientific Monthly* that great care must be exercised in assessing their evidence. (A Scottish court decided, not long ago, that a police dog which sniffed out a suspected house-breaker, was a 'credible witness'.) Just how essential such care is can be appreciated when it is pointed out that the author (a doctor of veterinary medicine) of a modern (1955) American book on the blood-hound makes the following rash statement: "Blood-hounds are not needed to recover wounded game by their blood. A man can follow that with his eyes. Any cur can be taught to follow blood odors, but neither a man nor a cur, nor any invention of science can tell that a certain individual wearing a leather shoe, or going barefoot, walked down a hard surfaced path three days before. A blood-hound often can." And the same author claims that blood-hounds commonly follow trails ten to twenty-five miles long. They

[1] More than 100 years ago, Giannuzzi found that dogs can apparently "smell" their food *after* section or destruction of their olfactory nerves.

have even, he says, been successful with much longer trails (*e. g.*, 135 miles), recording one which, he affirms, was more than four days old.

It is, no doubt, because beliefs of this kind are prevalent that some writers have evolved wave theories of olfaction and that others have lapsed into frank occultism. Such a one was Rudolph who maintained that because a dog he watched correctly followed a weak human track, ignoring in the process fresh cross tracks of children, much strong-smelling horse dung and also much strong-smelling sawdust, it must have been guided by irradiations given out by living cells, absorbed by various materials and then re-emitted. (It will surprise nobody to learn that Rudolph's views were published in a journal entitled *Psyche, oder Übersinnliche Welt* and having the sub-title *Monatlich erscheinende Zeitschrift für den gesamten Okkultismus und alle Geheimwissenschaften, für wissenschaftliche Erforschung der okkulten Phänomene des Seelenlebens, ferner für „Indische Philosophie", Theosophie, Spiritualismus, wahre ethische Kultur, naturgemäße Lebensweise und Sozialreform.*)

Not less ridiculous were some of the views later expressed by De Danne at the World Kynological Congress in 1935. He also put forward a wave or radiation theory, suggested a process involving electrical energy and spoke of using a pendulum to determine olfactory acuity. More sensibly, he maintained that dogs can be trained to follow the tracks (as much as eight or ten hours old) of wild but to ignore those of domestic animals and that domestication has had a detrimental effect on dogs, causing their sensory powers to diminish.

And even in 1957, Copelman hinted that dogs have a peculiar occult power to distinguish by olfaction between normal adults on the one hand and dwarfs or children on the other and that, in general, they fear children and dwarfs or are jealous of them.

Observations of the behaviour of puppies and tests in which mint, camphor and ammonia were used have shown that their olfactory powers are considerable even at birth and rapidly develop as they grow. First they smell their mother's milk at increasing distances, later, their litter-mates, then food and food containers, then their master, after that the urine of adult dogs and, finally, more or less everything. When they reach full maturity, their olfactory apparatus has acquired the size and structure which accompany great acuity of sense [2]. This acuity, perhaps, is much superior to that of man whose olfactory region is much smaller. However, it (and also the area of the olfactory region) varies appreciably from breed to breed and can be increased, possibly greatly, by breeding (especially in the female line) and training, hereditary influences perhaps playing a part.

---

[2] Even for flower odours to which they may normally show aversion or indifference.

Dogs exhibit, in addition to variations in acuity, great differences in ability or willingness to apply their olfactory powers in tracking, a fact which is of great importance to those responsible for the selection and training of trackers. No less important is the selection of the trainers themselves and it has been repeatedly observed that tracking powers only reach their full development after long, intimate and completely sympathetic relationship between the dog and his trainer, who will frequently be also his master, has been established. The good trainer will, of course, ensure that his dog is always well nourished and in perfect health. He will remember, too, that the animal will use (and sometimes, perhaps, will be diverted or led astray by) its senses of sight and hearing.

The canine ear, it seems certain, is very sensitive and can be of great value, especially in the dark. The eye, on the other hand, is apparently much less sensitive. Indeed, according to the experiments of Schmid, dogs cannot detect objects by vision at distances greater than about 600—800 metres and cannot recognize if the distance exceeds about 150 metres. (The distances given by Vesey-Fitzgerald are much smaller.) The visual power, of course, depends on the size and breed of dog as well as on other factors the most important of which, perhaps, is whether the objects are stationary or moving.

Perhaps the most fascinating aspect of the relationship between man and dog is the ability of the latter to recognize and track the former by scent. But before this aspect is considered, it seems desirable first to review the discussions and laboratory-type experiments on this ability which have been conducted on several occasions.

As early as 1872, numerous references to olfaction in dogs (and other animals) appeared in *Nature* when Darwin, Wallace and others took part in a discussion of 'inherited instinct' or 'ancestral memory'. Wallace suggested that a dislike by dogs of the flesh odour which clings to butchers may be inherited. Homing, he believed, is probably also an olfactory phenomenon. Animals shut up in a basket inside a coach possibly find their way home by remembering a series of linked odours and returning by the inverse series. (One correspondent reported that a dog taken nearly 1000 miles by rail from Mentone to Vienna found its way back to Mentone and, many years later it was reported from Canada that an Airedale found its way home over a 170-mile stretch in six weeks.) A proposal that homing tests should be made with cats shut up in a basket and that stale fish should be placed in the basket so that its odour would mask external smells was apparently not adopted and, in any case, was regarded by some as anthropomorphic. Tests with blind dogs were also suggested and Darwin told of a blind horse which regularly stopped outside inns which it probably detected by smell. It was pointed out, too, that smell is to dogs what sight is to man and that man can scarcely appreciate the importance of smell to animals, especially dogs. However, there seems to be no evidence for the view that man or beast can learn an 'odour series' apart from the objects emitting the odours, *i. e.*, perceiving the odours but not seeing the objects. Associations are between odours and objects, it was said, not between odours and odours and it was remarked that in non-human animals a very important factor is exclusive direction of attention of a sense as when birds apparently hear hidden worms despite much louder adjacent noise. Moreover, odours are notoriously transient and homing animals frequently return by a route different from the outward. Concluding the debate in *Nature,* Croom Robertson argued that "it had never been sufficiently considered what help in finding their way dogs might have from smell alone. Be that help what it may in the particular cases, I thought it clear that if in their common experience smell does not somehow supply to dogs the defect of touch they are, as far as we can see, badly fitted out, by comparison with men, for making their way through the world. And . . . I must still . . . hope that the notion of a continuous world of smells is not an impossible one. If the external world were the same to dogs as it is to men — a complex of interwoven touches and sights in space, and only in addition dogs had more frequent and varied experiences of smell, the dying away or shifting of some in a particular train of odours would doubtless . . . put a dog out when reduced to work its way back along such a train. But my point is, that a dog will regularly think of all things, stationary or moving, by their smell, where we think of them by their touch as handled, and upon this simple ground of fact that a dog has no hands; in which case the continuity of a road will as little to the dog as to us depend upon the standing-still of flocks of sheep or any other passing objects. It is true that our experience does not enable us easily to fancy what sort of world this of the dog's will be, but at the worst, we need not conceive it . . . upon the analogy of a succession of coloured mists. Do we, even with our intermittent smell, find it so impossible to refer the diffused odour of a dunghill to a particular source? . . . Of course, no train of evanescent smells can guide a dog back upon a road from which they have died away; as little, or still less, can the succession of particular smells, however constant, lead a dog upon a line that he has never travelled over before. But that dogs, while they have no such touch as ours, do constantly use their sense of smell to guide them cannot be doubted; and the result to them must be such a very different world of experience from ours, though developed under common laws of acquisition, that we have no means of deciding what it is impossible to be done by *some* dogs through mere experience."

These important discussions in *Nature* were soon followed by the interesting essays on smell in vertebrates published by Grant Allen in 1880 and 1881. Allen, who included a very informative discussion of the sense of smell in his book *Physiological Aesthetics,* believed that, as we rise in the intellectual scale, sight increases and smell decreases. Dogs cannot *handle* the objects in which they are interested, he observed, so they *smell* them and their sense of smell is of immense value to them as it is to most of the lower animals. "The bloodhound", he somewhat incautiously asserted, "can track a human scent a week old, which argues a delicacy of the nose almost incredible to human nostrils". And he added that dogs even *dream* about smells.

The observations of Passy and his collaborator Binet, made nearly twenty years later, were of a different kind and, perhaps, less interesting. Moreover, their results were obtained with one dog only (a pug), four others having proved unsatisfactory. Paper dipped in an alcoholic solution of odorous substance was allowed to dry and then presented to this animal. The odours of heliotropin, natural and artificial musk, violets, vanillin, *iso*valeric acid, lauric acid, caproic acid, camphor, roses, santal and castoreum pleased him, he smelled them attentively, seized the paper and attempted to eat it. Odours which displeased him were those of orange, jasmine, alcohol, ether, geranium, the displeasure being shown by turning his head away from the paper and looking at it from the side "with a comical expression of disdain". As would be expected, he preferred animal odours to others. It seemed that he readily experienced olfactory adaptation (fatigue), rejecting at a second test some odours found pleasing at the first. The unpleasing odours were never accepted at a second or later test. This dog was also able to distinguish natural from artificial musk. Passy and Binet made some tests with asses, cows, goats, horses and sheep. They found that sheep and, even more goats, apparently dislike the odour of tobacco and that of human breath and will reject leaves bearing these odours even when thoroughly mixed with many others not so tainted. But their experimental methods appear to have been rather primitive and the numbers of their subjects much too small.

About ten years were again to elapse before another test of the olfactory power of the dog was made, this time in America, by Hamilton. But although he used a well designed apparatus he also tested only one dog and his primary interest was to discover whether it could learn to distinguish colours (black, green, red, yellow). He also used, alone or as 'reinforcements' for the colours, odorous substances (asafoetida, lupulin, castor oil, beef extract). However, he was obliged to conclude that it was impossible to tell with certainty whether the dog was ever guided by the colours and that the reinforcement with odours seemed to have little value.

In Russia, about this time, several physiologists applied the method of conditioned reflexes to the study of olfaction and others senses in dogs. As odorous substances they used amyl acetate, asafoetida, bergamot oil, camphor, clove oil, meat, mint, naphthalene, tributyrin, turpentine and vanillin, and they were able to show clearly that the animals could distinguish between odours in which they presumably could have no biological interest. In some cases, their dogs were first blindfolded or subjected to brain operations.

Somewhat similar experiments, with similar results, were carried out in the U.S.A. during the period 1937—1943 by W. F. Allen who used the same and other (acetic acid, alcohol, ammonia, anise, benzol, butyric acid, chloroform, ether, eucalyptus, formaldehyde, lavender, phenol, pyridine, wintergreen and xylol) odorous materials. Allen showed, also, that dogs, like human beings, apparently have trigeminal as well as olfactory sensations.

In Germany also, experiments on the olfactory capacity of dogs were begun about this time. Working in Berlin, Kalischer set out to train dogs to take food only when it was associated with a specific odour and not otherwise and also to find to what extent they could detect such an odour when mixed with other odours. His dogs soon showed that they were able to do both these things. Thus, they could detect the odour of benzaldehyde even when this compound was mixed with an excess of the very similar-smelling nitrobenzene or with an excess of eau de Cologne, although, in both cases, the benzaldehyde odour was rendered completely indetectable to man. One dog, trained to *iso*valeric acid, could detect it when mixed with acetic, propionic, butyric and caproic acids and another, trained to natural musk, could detect it in admixture with artificial musk. After an interval of five months, the animal still retained this power. Kalischer found that male dogs learned better than did females. He obtained no evidence that his dogs suffered from olfactory fatigue (adaptation).

The investigations of Heitzenroeder were carried out in Giessen a few years after those of Kalischer. He endeavoured to avoid the dangers of subjectivity by using, as a measure of olfactory response, kymograph records of the duration and intensity of the respiratory movements (sniffing) induced by odours from each of Zwaardemaker's nine classes and by other materials. (This procedure was also adopted by Srnetz in 1935 in studying olfaction in the horse, dog and other animals.) Since the tests were of a preliminary character, he used but one dog, a male Pomeranian aged eight years. Its snout was introduced into a funnel to which was delivered a current of air containing the odour tested and round its body a belt, connected to the kymograph, was fixed. Heitzenroeder found — not unexpectedly! — that the animal reacted vigorously to the odours of bitches' urine and roast fowl. But he also found at least some reaction to practically all the odours employed. (Other substances used[3] were: camphor, carbon bisulphide, coumarin, eucalyptol, ethyl mercaptan, heliotropin, musk, naphthalene, rubber, skatole, vanillin and yellow wax.) Little or no response was elicited, however, by the odours of violets, roses and stocks. Fatigue was observed to occur readily.

Heitzenroeder's dog did not cease to serve the cause of science when the experiments just described were completed. After an interval of about

[3] In concentrations just discernable by man.

two years, he was the subject of a series of tests by Seffrin, who, using the same technique as did Heitzenroeder, sought to determine some olfactory 'minima perceptible' to this animal. Seffrin also used some pure substances from Zwaardemaker's classes as well as a number of animal products such as the flesh of the rabbit and the urine of a bitch. The numerical values obtained need not be quoted here. If they have any significance at all, they can scarcely be regarded as more than very rough approximations. But he did reach a very important, though as we shall see, controversial, conclusion. "Man", he said, "is superior to the dog in the detection of various pure odorous substances", although, of course, "the dog is much superior to man in the detection of the not accurately defined odorous substances of animal origin ...".

It was not long, only two or three years in fact, until the procedure of Heitzenroeder and Seffrin and, in consequence, some of their principal conclusions were challenged by Henning. Henning's experimental method was not very satisfactory, but later work has confirmed his views. He found that dogs can select from a number of handkerchiefs that one which alone is faintly perfumed with a flower odour and, conversely, that they can be trained to ignore even powerful animal odours which normally elicit very lively responses. He concluded that methods based on motor response teach us nothing about the range or acuteness of the olfactory powers of animals since absence of response does not necessarily indicate absence of ability to perceive. Moreover, he maintained that the dog, far from being unable to perceive non-animal odours (though it is normally indifferent to them), is not much inferior to man in its power to perceive them. In its 'olfactory continuum', he claimed, there are no gaps.

The researches so far discussed seem to have been unknown to Johnson of the American General Electric Co., when, in 1914, he published his particularly interesting '*Note on the Supposed Olfactory Hunting Responses of the Dog*'. Impressed by the facts that the scientific study of canine olfaction had been avoided (possibly because the olfactory stimuli could not be measured or controlled) and that the only data he had encountered were "meager, conflicting, untrustworthy and purely anecdotal" he considered the problem on the basis of the assumptions that a fox-hound can follow a three-hour-old trail of a rabbit without back-tracking and that this is very near the limit of the dog's ability. Hunters maintain that the dog follows the toe-line because the more recently made tracks have a stronger odour than those made earlier. But, argued Johnson, consideration of gaseous diffusion processes and rates of decay reveals that, if the amount of odorous substance deposited in each set of foot-prints is the same and if the time interval between the sets is one second, then the intensity of the scent in the track would only increase if the dog were to travel faster than the rabbit did and, even then, the increase would be very small indeed. Moreover, if the track were to pass from moist ground to dry ground, the stimulus intensity could be less on the recently made part of the track than on the part made earlier. Something like this may also be the case if the odorous substance is deposited in liquid or solid form, if decomposition of the substance occurs or if other complex changes take place. For these and other reasons, Johnson doubted the intensity-difference theory and was inclined to favour Cobb's suggestion "that the dog's sense of direction may be due to the trailing of *ground* smell-substances. For instance, the smell-substances affecting a dog trailing a man who had crossed a mint bed might be: 1. ground+man; 2. ground+man+mint, the mint being intense; 3. ground+man+mint, the mint smell-substances diminishing rapidly in the direction the man had taken." While regarding this as a valuable although incomplete explanation of the dog's tracking power, Johnson also asked whether the tracking behaviour of the dog is really an olfactory response at all and he proposed that a comparison should be made of the field behaviour of anosmic and normal dogs of the same litter and of a hunting breed and that other experimental investigations should be conducted.

After the first world war, the German interest in canine tracking powers which had led to the experiments of Heitzenroeder, Seffrin, Henning and others continued not only in police and hunting circles but also amongst students of physiology and animal behaviour. The huntsmen, recalling the fact that the nose is only the receptor whilst perception takes place in the brain, also discussed the confusion which had arisen in consequence of the use of erroneous terminology and the misuse of words. Some of them, too, emphasized the difference between 'direct scenters' (with head up) and 'indirect' (with nose close to ground) claiming that these types exhibit corresponding anatomical differences and that, although olfactory acuity can be improved by selective breeding, some direct scenters cannot perform indirect tracking. There were, however, considerable differences of opinion amongst the huntsmen and, in this respect they resembled physiologists and psychologists such as Lutz, Pfungst and Vogt who devoted their attention to canine behaviour in the early post-war years. In a very valuable general contribution, Lutz stated that the dogs which successfully find their way home from a distance are only a small proportion of the total and that the problem of homing is merely one of probability! No dog, he said, can follow a track more than a few hours old and no dog had so far been proved completely track-faithful. Dogs learn by association and by forming conditioned reflexes, not by thinking (or, at most, by very primitive thinking). When tests of their tracking power are being conducted, the greatest possible care must be taken to exclude sources of error and handlers must never know in advance the solutions to the problems set. In training and tests, Pfungst recommended the use of artificial tracks laid with chemicals applied with dropping bottles. On these tracks dogs learn much more rapidly and go much further than on natural tracks. He remarked that the influence of meteorological conditions and especially of air currents at low levels should be carefully studied and he considered the wider adoption of some of the procedures used in the training of military dogs in rescue work. Commenting on the experiments [4] of Friedo Schmidt, he rejected the claim that penetration of new boots carrying artificial scent by human individual scent had been demonstrated and showed that the dogs did not cease to follow the track even when a second man laid the continuation of it.

Although it is improbable that the dog can appreciate human purposes, it must, nevertheless have some conceptions and we have no right, argued Vogt, to deny it power to think and to maintain, as Most did, that it learns entirely by memory of purely external space and time relations. The dog's ability to recognize sounds and to perceive minute movements (*e. g.*, of human facial expression) must not be forgotten.

Vogt's views did not escape criticism and tests which he conducted at Grünheide were held to be inadequate by Most and others who maintained that, in practice, some experimental results were not confirmed and that a test with a dog's master need not yield the same result as a test with a stranger.

If a dog is to succeed in tracking human beings, it must, it would seem, be able to perceive a given olfactory impression which preserves some degree of constancy, to retain this in its memory for some period, long or short, and in some way to show that it recognizes a second olfactory impression as very similar to or identical with the first. If the individual odour theory is sound, this odour is the constant guiding factor and it is the factor to which the dog must continuously attend. It is commonly assumed that dogs, even when untrained, can do this and, indeed, that the fact has frequently been demonstrated in criminal procedure. „Dem ist entgegenzuhalten daß exakte Beweise für obengenannte Fähigkeiten bisher noch nicht erbracht werden konnten und daß alle Hundeuntersuchungen nach dieser Richtung unter streng wissenschaftlichen Bedingungen zu negativen Ergebnissen geführt haben." This conclusion, which may be briefly rendered "it has not yet been proved", was expressed by Löhner, professor of physiology in Graz, in 1927. He began his experiments in Bonn in 1912. Several years later, he studied biochemical specificity and individuality with special reference to the secretion of odorous substances. He then showed that human beings — and presumably also dogs, though this was not actually shown — can distinguish in man both 'regional' and 'individual' odours, the individual odours being supposedly simply mixtures of the regional odours which themselves emanate from various parts of the body, in particular from those which are covered with hair. All these odours belong to Zwaardemaker's

[4] These experiments are described below.

'caprylic' class. Löhner endeavoured, as far as possible, to eliminate the many sources of error involved in studies of this kind. His subjects were healthy men, twenty to forty years of age, some of whom were on identical diets from which certain foods such as onions were excluded. Care was taken to minimize the effects of perspiration and of 'decomposition' odours. The subjects washed themselves thoroughly using soap free from perfume. It was found in these experiments in which pieces of cloth were fastened to various parts of the human body for periods of fifteen minutes or less before being smelled that, to man, the differences between the regional odours of one and the same person were greater than the difference between the individual odours of two persons. It was also shown, as would of course be expected, that scented objects kept in closed containers retained their scent longer than did those exposed to the air and that increase of temperature decreased persistence of scent. In the tracking of barefoot men (ignoring cross tracks) and in the retrieving of pieces of cloth to which human scent adhered, some of Löhner's well trained dogs showed great skill. An experiment with a goat, however, which was found to give surprisingly correct 'answers' when its master knew them but failed badly when he did not strengthened Löhner's opinion, formed after studying the literature on the tracking powers of the police dog, that additional investigation was required. Accordingly, he decided to institute experiments in which disturbing factors could be controlled or excluded to the greatest possible extent. Carefully cleaned sticks having the dimensions $100 \times 15 \times 15$ mm., deodorized at 150° C, for ten minutes, were used. The dog's task was to take the scent (thoroughly sniffing) usually from the carefully cleaned palm of the hand of the observer, to select from amongst ten to twenty control sticks that piece which had been touched or handled by the observer (or, in some cases, had been in contact with some other part of his body) and deposited at the end of a track and to fetch the piece to him. Special attention was paid to the psychic condition of the dog. She was not required to cooperate when tired or when she had lost interest and she was rewarded with a tasty morsel when each task was successfully completed. The results of these tests were not less astonishing than were those of Romanes which will be described later. Even when the stick had been in contact with the finger tip for not less than two minutes, the scented side being downwards, the dog retrieved it successfully. When the whole palm of the hand was used to transmit the scent, the duration of contact did not need to exceed one or two seconds. Moreover, odorous liquids such as alcohol and strong-smelling essential oils — oil of cloves, for example — applied to the wood or the hand of the observer, did not prevent the dog from choosing correctly though she clearly found some of the odours very irritating. She also was not diverted from her purpose when the piece of wood, scented by the observer, was then handled by four other persons or when the observer's piece of wood was handled by other persons before he handled it. When she had to choose between two pieces, one of which was handled first by the observer and then by another person, the other being handled first by the other person and then by the observer, she always made, as her first choice, the piece last handled by the observer. After sniffing at the *hand,* this dog successfully retrieved sticks which had been in contact with *other parts of the body* than the hand but no conclusive proof was provided that she distinguished human regional odours.

Löhner pointed out that these experiments must be regarded as laboratory investigations and he expressly refrained from giving judgment on field tracking. If such tracking is to be trustworthy, the dog must be able, from the scent of an object or track, to distinguish the person from whom the scent has been transmitted, and, conversely, to distinguish an object or track after having taken the scent from the person whose scent has been transmitted to that object or track. It must be able to apply its power, not only to its master or trainer, but also to strangers and it must be capable of ignoring other scented objects when the object with the scent to be traced is absent. In an attempt to provide evidence that the dog is indeed able to perform such tasks — after training, if necessary — Löhner conducted still another series of tests with the dog "Bella" which had already served him so well. Five tests in all were made, each test (other than the first) being instituted only after the task of the preceding one had been mastered. As before, sticks were employed. Bella had to take the scent at her master's hand, follow a track which he laid and retrieve the stick which he had deposited at the end of the track. She had then to repeat this task after three other persons had likewise laid tracks radiating from the point where her master's track began and crossing his track and the other tracks at several points. The other persons, of course, also deposited sticks. In the next test, sticks were not deposited at the ends of the tracks but were thrown forward for increasing distances. The master's stick was used in the fourth test but it was carried by another person who, consequently, laid the track. In all these tests, the tracks were, apparently, rather short (about 25 metres). Finally, four sticks, each handled by one person only, were deposited with four control sticks on glass plates in an enclosed space. These sticks were 10 cm apart and were deposited with long tongs by an assistant in the absence of the dog and of the persons taking part. Once again, Bella had to take the scent at the hand of one person and then retrieve his stick, being rewarded or scolded according to the result. When, three times in succession, she had successfully retrieved the stick of one particular person, he withdrew from the test and his stick was replaced by a control stick. After these tests had been in progress for several weeks, Bella's skill had so improved that she was frequently 100% successful in this final test and Löhner was inclined (at least with regard to the retrieval of objects in the absence of those of the dog's master, no track-following being involved) to adopt an opinion somewhat less negative than that quoted above. "Der Hund", he said, "ist ein Geruchstier, sein Weltbild ein Geruchsbild". However, when he made a re-evaluation of his results, taking into account the fact that the sticks were not cleaned after being correctly retrieved so that the first and second repetitions were made, in each case, with sticks which retained material from the dog's mouth, he felt obliged to remain doubtful. The psyche of the dog, he pointed out, is very labile. Slight variations in customary procedure by the trainer suffice, occasionally, to cause complete failure for a whole day. Above all, the complexity of the problem is very great and extreme reserve must be maintained in coming to conclusions.

The problem of identifying *strangers* by scent and of indicating identities of scent is, Löhner believed, partly physiological, partly psychological. Physiologically, all conditions necessary for the solution of the problem seem to be satisfied both as to individual differences in human scent and to the capability of the dog to distinguish regional and individual human scents. These are not suppressed for the dog by odours which appear very much stronger to man. And the scents are adequately transmitted to objects within a very short period (one or two seconds). The adequacy of all the steps in the chain, except the giving of the scent and its acceptance by the dog had frequently been demonstrated. But can a dog familiarize itself rapidly enough with the scent of a previously unknown stranger? May it not require nearly a whole day in order to accomplish this? And must not a criminal who is to be tracked almost inevitably be a stranger? Whatever the answers, it must never be forgotten that the canine species has undergone very profound changes in the course of domestication and that successful training must utilize and advance the dog's own interests, not attempting merely to make him an instrument in man's hands and taking account of the fact that specialization is essential for highest achievement. Thus, the dog trained as a tracker must never be employed for any other work. It was attention to points such as these that enabled Most to devise exact, speedy and successful methods at the German Army Canine Establishment at Kummersdorf (near Berlin).

Löhner[6], it seems, conducted no olfactometric tests on dogs nor did he compare the olfactory acuities of various breeds. This was attempted, however, by Hoeft and by Ströse in a rather primitive way. They used varying concentrations of meat extract in water to which colouring matter was added to eliminate a visual clue. Coloured water served as control liquid. When the dogs (spaniels, Alsatians, dobermanns and others) lapped solution in preference to water, a positive response was recorded. More than

[6] A brief review of Löhner's work was published in 1926 by Ide in the *Revue Médicale de Louvain.*

twenty years later, the Russian investigators Krushinskii, Chuvaev and Volkind also described a simple method of testing canine olfactory power. They found greatest acuity in Laikas and Alsatians and observed that males were superior to females. And they also showed that acuity can be increased (thresholds lowered) by training.

Precise measurements were first made, however, in the period 1953 to 1958 by Neuhaus whose careful and exact work, although not directly in the domain of tracking, has far surpassed that of any other investigator in this field. He made more than five thousand tests, many of them with one dog (a fox-terrier bitch) using an apparatus designed by himself. In this apparatus, purified odorous or odourless air was delivered to three points at which the dog made a choice and was rewarded by gaining access to a lump of sugar when the correct choice was made but not otherwise. All the odorous substances used were liquids or dissolved solids contained in a capillary tube the contents of which were measured microscopically as they were mixed with air before passing to the point of delivery. Errors due to adsorption were minimized by allowing equilibrium to be attained and changes in the concentration of odorous substance as small as one part in three million parts of air could be made. Most of the substances could be assumed to be of biological importance to dogs. (They included lower fatty acids, especially butyric acid, and their esters, diacetyl and ethyl mercaptan.) But other substances, such as α- and β-ionone, eugenol and nerol, were also used. (The dog's power to detect these 'flowery' odours also much surpassed that of man.) In some cases, sausage was given as reward instead of sugar when the correct choice was made but, because of the meaty odour of this material, it was not released in the apparatus. The psychologically (and probably also physiologically) disturbing technique of Heitzenroeder and Seffrin was avoided, the dog being free to sniff at will at the points of delivery, there distinguishing odorous from odourless air, more odorous from less odorous or one odour or combination of odours from another. Neuhaus also made histological examinations (vital staining) of the canine olfactory region, finding that, according to the breed of dog concerned, it contained between 125 million and 22.5 million cells. These cells appeared to be of four types but, since intermediate forms were also observed, there may actually be as many as forty or more. He noted that the dog, like man, exhibits fatigue (adaption) when exposed to high concentrations of odour for more than a short time and that it also has, apparently, a threshold of *perception* which, of course, is lower than its threshold of *recognition*. He found no evidence that neutralization (compensation) of odours, sometimes said to occur for man, occurs for dogs.

The threshold values found by Neuhaus were not simply low, they were almost unbelieveably so, ranging, in the fatty acid series from 9000 to 50,000 molecules of substance per cubic centimetre of air. (For the 'non-biological' odours, the values were not very much higher.) Moreover, by using two-component, four-component and even five-component mixtures, he was able to show that mixtures of like odours in which each and every component is present in sub-liminal concentration may still be perceptible to the dog because some kind of synergistic process occurs. Thus, the concentration of caproic acid which must be present in order to enable a dog to distinguish between two four-component mixtures is *less than* the threshold concentration of that acid when it is presented alone. And, in the same way, a dog can perceive the odour of a two-component mixture in which one odorous substance is present in a concentration just sub-liminal and the other in a concentration much below the threshold value. Neuhaus included in his studies some experiments on the effects on canine olfactory thresholds of oral administration of odorous substances, finding that when the odour of the substance taken by mouth is like that of the substance smelled, the threshold is first raised , then falls below normal and finally returns slowly to normal. He believes that the material ingested may possibly activate enzymes in the olfactory mucous membrane at the stage where the threshold is lowered. Also, he considers it probable that, in the dog, groups of olfactory cells may respond preferentially (but not with any high degree of specificity) to certain classes of odours and he calculates that, quite possibly, one molecule of odorous substance may be enough to stimulate one cell. If these assumptions are correct, and if the number of odour classes is not large (say, one hundred or not much more) then the number of possible combinations of odours which could be perceived would be very large (exceeding ten thousand). In experiments with varying concentrations of the same substance, he found that differences of 60% and 80% were perceived but that those of 50% were not and he concluded that these results provided a rough confirmation of the Weber-Fechner law [7]. Some of his observations, too, led him to incline to the view that strong odorous power is correlated with solubility in water *and* in oil.

As a consequence of his investigations, Neuhaus was naturally able to express opinions which must be considered very weighty on the tracking of human beings by dogs. He had shown that, since enough odorous material diffuses through ordinary footwear, human beings can themselves smell the imprint of human feet on previously odourless paper, having first proved that this was theoretically possible. Human perspiration (it might be, say, a ten-component mixture) is probably always much the same qualitatively but varies quantitatively *(i. e.,* exhibits differences in the proportions of the various constituents). Even if the concentrations of *all* the odorous constituents are very subliminal, the odour might still be detectable in a track without any of the 'summation' supposed by the Menzels to occur. In the

---

[7] See p. 107.

human track, the odour from the foot is certainly perceptible to the dog even when other and, to man, much stronger odours are also present. Moreover, since the dog can detect differences as small as 0.1% in total concentration and since it can presumably derive clues from the changes produced by the different rates of evaporation of the various odorous materials in the track, it can be regarded as entirely certain that track-fidelity in dogs is a possibility, especially for the breeds, such as Alsatians, which have superior olfactory acuity. (The acuity of the Alsatian was found to be about 35% greater than that of the fox-terrier.) Having an olfactory acuity which is from one million to one hundred million times that of man, a dog can detect one milligramme of butyric acid in one hundred million cubic metres of air, the 'volume', said Neuhaus, of a whole town. And, summing up his conclusions, he wrote: "Es folgt daraus, dass dem Hund mit seiner für Buttersäure z. B. millionenfach besseren Riechleistung über viele Stunden hinweg eine überschwellige Konzentration in einer Fuss-spur des Menschen dargeboten wird. Damit wird die Spurenreinheit mancher Hunde, d. h. die sichere Unterscheidung gleichalter Spuren verschiedener Menschen und Tiere, physiologisch verständlich, wenn man weiter berücksichtigt, dass die einzelnen Komponenten im Mischduft des Schweisses individuell abgestuft sein dürften."

While the work of Neuhaus was in progress, a joint British-American research project concerning canine olfactory power was instituted. In the American experiments, conducted by Becker, Markee and King at Duke University, North Carolina, thirty-six dogs were tested in three types of problem cages, using asafoetida, bouillon, clove oil and paraffin, the object being to establish olfactory discrimination. But when the dogs made a successful response, it was usually doubtful whether or not they used olfactory clues. The investigators, who seem to have ignored the work of many of their predecessors (including much of the work of "Konrad Most, a Berlin police investigator"), did not record the types of dog tested. Their animals were closely confined and obviously anxious to escape and, the design of the experiments being poor, it is not surprising that the results were disappointing.

Ashton, Eayrs and Moulton undertook the British part of the research at the University of Birmingham. Their experimental technique was much superior to that of the Americans and they conducted a very large number of tests but with two dogs (cross bred Labradors) only. The object of their tests was to determine the thresholds for eight straight-chain fatty acids from formic to caprylic. They used a system of light signals and bells in a laboratory twelve feet square with two ante-rooms, arranging the procedure so that it was impossible for the supervisor (in one ante-room) and the dog's handler (in the other) to give clues to each other or to the dogs. Further, they ensured that no human scent was transmitted to the apparatus used. They found that the threshold concentrations tended to vary inversely with the number of carbon atoms in the molecules of the odorous acids used. After subjecting their results to very thorough statistical examination, they concluded that the olfactory sensitivity of the dog is not very different from that of man. However, they admitted that a much larger number of animals should be tested. They did not convert the values (molar concentrations) they obtained into the corresponding numbers of molecules per cubic centimetre of air but this was done (for butyric, valeric, caproic and oenanthic acids), two years later, by Davies and Taylor with the surprising result that the values obtained were ten million times greater than those of Neuhaus and Buytendijk [8]: In consequence, Davies and Taylor concluded either that the tests of Ashton and coworkers were not conducted under equilibrium conditions or that their dogs were less sensitive to odours than had been supposed. Perhaps it is more probable, however, that this enormous difference may have been due chiefly to the much lower degree of precision of the method of the British workers [9] in whose experimental technique the various concentrations of odorous solutions were simply exposed in crucibles at which the dogs sniffed.

A low degree of precision, however, did not prevent the Russion investigator Myznikov from obtaining results which appear to be more in harmony with those of Neuhaus than with those of Ashton and his colleagues. Myznikov had worked for four years with police dogs in natural working conditions and found that they could detect scent one day old or more. Quoting Frolov, he stated that whereas man can distinguish a few thousand smells only, dogs can easily distinguish about half-a-million. In his experiments, he employed thirty East European sheep dogs, using aqueous solutions of ammonia, acetic acid and thymol and making more than three hundred tests. The solutions, together with distilled water as control material, were poured into small bottles which were wrapped in cotton wool and cloth and placed at 15—20 cm intervals on a wooden stand. The dogs, sometimes blindfolded, were allowed to sniff at gauze moistened with odorous solution and then caused to select from the stand, which was placed at a distance of three metres, the bottle having the odour concerned. The time taken for selection, the wind velocity, the air temperature and the relative humidity were noted. The lowest concentrations selected were regarded as the threshold concentrations. The values (presumably in grammes per litre) were: ammonia, 10^{-13} to 10^{-24}; acetic acid, 10^{-13} to 10^{-21}; and thymol, 5×10^{-14} to 10^{-20}. Myznikov noted that, amongst his dogs, considerable differences in olfactory power existed, these differences being much greater than those between the thresholds of the same dog for the three substances. He also

[8] See p. 43.

[9] Lower than that of Neuhaus.

observed that the dogs which could follow the oldest tracks were those having the highest acuity. A particularly interesting discovery which he recorded was that acuity was greatly increased (for ammonia, about thousandfold) by administering caffeine which likewise greatly improved tracking power, the duration and extent of track-following being doubled. Somewhat similar results were obtained by administering bromide and phenamine.

This stimulating power of drugs has been noted by other Russian workers who state, for example, that human olfactory acuity is increased by caffeine, ephedrine and strychnine. Tame dogs, these workers found, are inferior in acuity to wild dogs. Acuity can be increased by training but diminishes again if the power is not exercised. Training has both central and peripheral effects.

The olfactometric method of Krushinskii and his co-workers was first mentioned, without detail, in 1946. Somewhat more than a decade later they published the details. They assess the acuity from the number of layers of fabric through which an odour (meat odour) can be detected, two boxes containing the fabric but one only having fabric with this odour being offered to the animals. Small doses (0.01—0.02 g) of phenamine, administered to them were found to increase the acuity which was further increased synergistically by caffeine. Acuity was diminshed, however, if phenamine was given daily for a long time. The effect of the drugs became noticeable after 30—40 minutes and lasted for several hours.

As has already been remarked, it is undoubtedly in their work with the police that dogs display the most fascinating (and also the most puzzling) aspect of their olfactory power. However, this is not the only domain in which man has applied this power and it may be of interest first to consider briefly some other canine services in which it is concerned to some extent. Perhaps the most remarkable of these services are in gas-leak detection, in war work, in Alpine rescue work and, of course, in hunting.

On one occasion at least, dogs have been employed with pronounced success to detect leaks in mains conveying natural gas odorized with stenches added to the gas as warning agents and for facilitating leak detection. Such dogs, trained with a liquid having a stink similar to that of butyl mercaptan, were able to locate a few drops of a solution of this substances in an acre of ground, being more efficient for the purpose than the instruments which were also used.

It is widely known that dogs have been used in war as sentries, guards and messengers and for rescuing the wounded. They have also been employed with success as mine detectors since they can locate by smell the 'wounds' made in the ground when the mines are buried. In various countries, schools for war dogs have been established and accounts of the work of such dogs with the American, British, French and German forces have been published.

Various breeds of dogs have been used. In the second world war, the U.S. Forces used dogs of all large breeds in all theatres and all climates but eventually selected Alsatians as the most suitable. In Britain, Major Richardson, a pioneer in the use of military and police dogs, worked with them in many countries for fifty years or more. Unfortunately, his training methods seem to have been far from scientific and it is impossible properly to evaluate the records which he published.

In the Alps, particularly at the St. Bernhard Monastery, as everybody knows, and elsewhere rescue dogs have frequently performed valuable life-saving service although many, perhaps the great majority, of the popular stories about the St Bernhard dogs have no foundation in fact. In any case, at the monastery in 1953, the dog trainer Heyer was told that "Les chiens ne travaillent plus". However, considerable numbers of dogs (Lawinensuchhunde) and leaders are now trained in Switzerland and in Austria for rescuing people buried by avalanches. Heyer has recorded one case where, of eighteen buried boys, seventeen were saved by a dog which barked and dug at the spot and another in which, in 1944, a man buried one and a half metres deep and a woman buried five metres deep were saved by a dog, the woman having been buried for 52 hours. It can hardly be doubted that these rescue dogs are guided by scent and, in fact, they are trained to search under the snow for objects bearing human scent. Heyer believes, as do the Russian investigators, that training improves olfactory acuity. Another expert who regards the individual odour (and perhaps also the so-called 'anxiety' odour) of the buried persons as the dog's guide is Hattinger of the Austrian Gendarmerie. He began training dogs for avalanche rescue work in 1950 and he reports a case where fourteen policemen took from one to five hours to find buried objects whereas a dog found the remaining object in five minutes, despite the fact that the place of search was covered with many human tracks. Search dogs, preferably Alsatians, he says, should be trained by carefully selected trainers in all weathers by day and by night (he has found that they work better by night) and in fog. Moving over the area in zig-zag lines, they should learn to search, first for objects, especially articles of clothing, belonging to their trainers, then for similar articles belonging to strangers. After that, they should practice locating and digging up food (meat) buried from one to two and a half metres deep. Finally, they should learn to locate and help to dig out, first their trainers and then strangers buried as much as five metres down under cover crossed by the tracks of many people. (Digging to this depth may occupy as much as one hour.) A curious fact observed by Hattinger is that the dogs seem first to scent the *heads* of buried people rather than the *feet*. It is worthy of note that Brückner, himself an expert in training dogs for rescue work of this kind, praises Hattinger's work highly although he shares with Most and Hansmann the view that dog's are *not* guided by individual odours.

Of all the useful activities of the dog depending to a very large extent on the exercise of olfactory power probably the oldest and best known is hunting and, of course, the dog is by nature a hunter. In the service of man, however, he needs to be trained from an early age onwards to locate and track those intact and wounded animals which are his master's quarry. The training, in which the difficulty of the tasks should be progressively increased, will resemble in some measure that given to dogs required for tracking human beings but, obviously, hunting dogs must learn to refrain from following human tracks. They will often follow blood tracks left by wounded animals and indeed animal blood as well as odorous chemicals are used in the laying of artificial tracks for training purposes. Such tracks are particularly useful since their whole length, strength, age and so on can be recorded, the conditions of the ground on which they are laid can be accurately described and the meteorological conditions prevailing can also be noted. It is said that artificial tracks of this kind were used as long ago as 1897 or earlier and even that experiments with men on stilts and with a track-laying wheel were conducted at that time.

Evidence that some modern huntsmen (fox hunters, especially) take an interest in the olfactory aspects of their dog's activities is provided by the publication of such books as Budgett's *Hunting by Scent* (1933) and Pollard's *Mystery of Scent* (1937). In his book and in a subsequent article in the *Field,* Budgett discussed at length the climatic, meteorological and other conditions which affect the nature and persistence of the scent of man and game (birds, wild boar, deer, fox, hare and others). He states that the most important scent-modifying factors are air-motion, humidity and temperature. The otter, he tells us, has a very keen sense of smell and has itself the most persistent scent of all the animals hunted in England. He adds that human scent has been detected (presumably not on the track but directly through the air) by bloodhounds at 370 yards and by wolves and bears at half-a-mile. Remarking on the persistence of scent, he records that a hound has been known to follow a track laid with animal perfumery fixatives, after 11—13 days. Although he at first thought that the 'body scent' theory was a myth, he later came to believe that, in ordinary tracks, the scent is derived from oily secretions deposited by the feet (the footwear, in the case of man). Some dogs, however, will follow a track produced in grass by dragging through it a heavy weight which emits no scent and the general 'body scent' is never effective as a guide except at very close quarters. With the scent indicators such as those which he and Pollard devised "it is possible to ascertain with certainty whether scenting conditions are favourable or unfavourable for hunting". Budgett, it should be remarked, repeated some of the experiments of Romanes (these are described later) and conducted some tests (use of stilts and of man suspended from a trolley wire, for example) similar to those of Continental workers whose investigations were apparently unknown to him. Pollard did not dispute Budgett's experimental results but rejected not a few of his conclusions.

"Bloodhounds", wrote an anonymous contributor to the *Lancet* recently, "detect colder scents than probably any other animal". They have a wider spectrum of olfactory stimuli and a more delicate sensory appreciation. This writer grovelled in vain on the ground trying to detect human scent but many hounds, he said, have hunted lines several days old and his own old bitch dug the kit of victims of a mountain accident from the snow after 47 days. The mental concentration required to follow one and the same track without changing imposes a noticeable strain which causes fatigue far in excess of that due to the physical effort needed. In view of this, it is not surprising that young dogs make mistakes until they acquire experience. When a fresh wind crosses the track, hounds run a line 10—40 yards parallel to the quarry's path. In the still calm of the evening, they will lift the head to mark a man a quarter of a mile off uphill. Although there is a genuine cadaver scent, personal scent probably lingers after death, says this author.

When the tracking of animals is considered, it must be remembered that many mammalian species (antelope, deer, elephant, llama, peccary, wart hog, for example) have scent glands on the feet or on some other part of the body. The odorous secretions of these glands naturally serve as guides to animals of the same and other species. The deer, it is said, has such glands on the feet, both in front and at the rear. And since the secretions of these smell differently, it is possible that they enable a following dog not merely to locate the track but to know its direction. Some observers, however, maintain that when a dog encounters a scent track, it does not immediately learn the direction taken by the quarry and is as likely to hunt the heel line as the toe line in the first place. Von Stephanitz however, maintained that a dog follows the toe line and not the heel line because the human and animal toe makes a deeper and more scented impression and not because each individual footprint in the forward direction has sufficiently increased odour.

Whatever the part played by scent originating in such glands, it seems just possible that another component of the track may be, according to Bühler, expired air. It will be recalled (Binet and Passy) that some animals dislike the odour of human breath. In warm-blooded animals, the expired air at the time of exhalation, is warm, moist, rich in carbon dioxide and, perhaps, odorous. Dogs are said to be able to scent the breathing holes which seals make in ice, even when these holes are covered with snow. Bühler suggests that, in this case, the scent can come from the expired air only and he proposes that tracker dogs might be put on the track of a person who passes all his expired air into a bag in order to discover whether or not this procedure would affect the ability to track. Alternatively, tests in which portions of stored expired air are released at intervals along a track might be instituted.

On a least one occasion, a scientific attempt to correlate meteorological and scent conditions has been made. For the purposes of this attempt, which was described by Pick in the *Meteorological Magazine* in 1922, it was assumed that factors favourable to keeping air-borne odorous material down to the level of hounds would be absence of upward air convection currents, absence of wind and high relative humidity. Such conditions would prevail, it was decided, if the air at ground level were cooler than that at two feet above ground, if odorous particles were hygroscopic and were consequently weighed down when humidity was high and if dispersion of the particles were prevented by absence of wind. During a period of six months, fox hunters reported on "the degree of scent prevailing on days on which hunting took place". Meanwhile, temperatures at ground level and at two feet above ground, relative humidities and strength of wind were recorded. It was concluded from the records of fifty-three hunting days that, in a general way, the degree of correlation was good.

Although it seems scarcely necessary to conduct experiments in order to show that "the social behaviour of dogs greatly depends on their olfaction", Freud and Uyldert, who reached that conclusion, did indeed carry out a few investigations into the micturation and copulation patterns of male and female dogs. In some cases they administered testosterone, in others they used cocaine to produce 'osmoanaesthesia'.

Much more valuable were the observations of Von Uexkull and Sarris who studied the canine 'olfactory field' (Duftfeld) noting precisely how dogs utilize natural and artificial objects as locations for deposition of urine and faeces and follow each other to these places using optical as well as olfactory clues. In their tests, they used odorous materials such as herring brine, lysol, skatole, human and horse urine and valerian. They found that skatole attracted little attention and that lysol was ignored.

Two other investigators, Beach and Gilmore, found, as would indeed be expected, that mature male dogs were attracted more by oestrus than by non-oestrus urine from bitches.

It may be remarked here that the false or, at least greatly exaggerated claims made for chlorophyll as a deodorant for human use, have also been made by American writers with reference to canine use. These writers, indeed, have asserted that bitches can be caused to lose some or all of their 'mating' odour by oral administration of this substance. Breeding capabilities, they say, are not affected.

Although dogs seem to have been used as trackers of human beings since an early period in history, it appears that, until the last decade of the XIXth century, systematic training for the purpose on any appreciable scale had not been undertaken. It was in Belgium, where the first dog show was held about the year 1890, that the initial steps were taken. At that show, most of the dogs were probably performing animals from the circus but, not long after, trackers appeared and those trained by the police of Ghent during the last years of the century soon acquired an almost world-wide reputation, some of them being exported to Britain, France, Germany and Russia. (The Ghent gendarmerie have now lost all faith in the value of dogs as trackers having found them untrustworthy on the few occasions on which they have been used in recent years. It is not all impossible, however, that the untrustworthiness was due to inadequate or incorrect training and to the difficulties caused by urban development.) In England, the *Association of Bloodhound Breeders* was founded in 1897 with, as one of its objects, the training of bloodhounds 'to hunt man'. According to Von Stephanitz, the German police were impressed by reports of the work of the Belgians dogs and suggested, in 1901, that tests should be instituted in Germany. Although this step was apparently not then taken, the German *Verein zur Förderung und Verwendung von Polizeihunden,* established in 1902 already had about 3000 members in 1911 and had launched its periodical publication *Der Polizeihund.* Even before this (in 1899) the *Schäferhundverein* had been founded in Germany, a society which played a very great and very important part in the study and training of tracker dogs. Meanwhile, large numbers of police forces in Germany and elsewhere *(e. g.,* Paris) had acquired patrol and guard dogs if not also trackers. According to Wagenknecht, the stimulus to the use of tracker dogs in Germany came from Professor Gross of Prague in the year 1896. As early as 1906, the Berlin police instituted tests for police dogs and, in 1908, dogs from Ghent were employed by the (British) North Eastern Railway Company at Hull. At one time, the German State Railways maintained a very large canine school at Roentgental in the neighbourhood of Berlin. It was in 1908, also, that a German police dog school was opened at Karlshorst where German dog handlers and, later, handlers from other countries attended courses. (This school, directed by Klein, was later transferred to Grünheide where Konrad Most became an early director. Cecil-Gurney, who was the first and perhaps the only Englishman who ever attended a course at this establishment, regarded it as by far the largest and most important canine school in the world. While he was there, Hansmann was the director. The courses for the training of dogs and their handlers lasted from three to four months, about two hundred dogs taking part each year. Most of the dogs were Alsatians but some were Airedales or other breeds. Some of the training took place at night.) Then, in 1909, Zell's pioneering work, *Der Polizeihund als Gehilfe der Strafrechtsorgane,* appeared throwing light on what he called a "von der Wissenschaft fast gänzlich vernachlässigtes Gebiet".

The neglect did not long continue and in the succeeding year, 1910, came the publication of one of the most interesting and valuable books on the subject yet written, namely Friedo Schmidt's *Verbrecherspur und Polizeihund.* Soon thereafter the works of Frey *(Der Polizeihund, seine Geschichte,*

Zucht und Abrichtung) and Wagenknecht *(Zucht und Dressur des Polizeihundes)* in Germany and Mathieu *(Le Chien Criminel)* and Huyghebaert *(Le Chien: Psychologie, Olfaction, Mécanisme de l'Odorat)* in Belgium were issued, the Belgian Shepherd Dog Association organized tracking tests and published rules for such tests, Konrad Most began his scientific investigations in Saarbrücken and the Prussian Minister of the Interior instituted official tests of German police dogs. According to the Belgian authors, Germany (and, perhaps, Britain) soon surpassed French-speaking countries in the training and use of tracker dogs. One British expert, Arundel, claimed that the dog is the most teachable of all animals, especially if its trainer has the right qualities. Bloodhounds, he said, are the best trackers: they can follow a track three days old.

After the 1914—18 war, Buytendijk (later the author of a remarkable book entitled *De Psychologie van den Hond)* (1932) conducted some very interesting experiments with dogs belonging to the Amsterdam police and, in Germany, Blunk described his successful training methods in his *Die Ausbildung des Hundes zur Spurenreinheit.* About this time, too, there was a large (140 dogs) police dog school near Pretoria in South Africa (said, perhaps erroneously, to be the only state-owned institution of its kind in the world). It was claimed that these South African dogs, which were said to run twelve yards to the leeward of the tracks, could pick up tracks as much as 72 hours old (in one case, as much as six days old). They were used for tracking human beings only and were never allowed to hunt game. Handlers were trained with their dogs at this school. In Egypt, the police achieved great success with their tracker dogs which began work in 1936.

After the second world war, the British Metropolitan Police, who had already employed dogs for a considerable period, constituted a new dog squad of Labrador retrievers but continued to use Alsatians also. Likewise, dogs were attached to the Devonshire Force in which bloodhounds are regarded as much inferior, as trackers, to other breeds. Popular accounts of modern methods of training and using British police dogs have been published in recent years by Holman (*My Dog Rex,* 1957) and Campion (*The Perfect Team,* 1959) but it seems that little or no strictly scientific investigation of canine tracking power has ever been undertaken in Britain (cf. Arundel: *Police Dogs and their Training,* 4th edition, 1948). In West Germany, the customs authorities now have a canine school, presumably established since 1945, at Oberdachstetten.

A review of the modern study of tracker dogs would be incomplete if it omitted reference to the controversy, which appears to have been bitter at times, between those who adopted and those who opposed the 'individual odour' theory. This controversy seems to have been particularly acute in Germany where there also was, it is believed, disagreement (or, at least, rivalry) between the police and the military kynologists.

The scientific study of canine tracking seems to have been initiated by Romanes and the account of his experiments, published in 1887 in *Nature* (and later in the *Journal of the Linnean Society* and the *Revue Scientifique)* may conveniently be taken as point of departure, especially as, in his opinion, they demonstrated "the almost supernatural capabilities of smell in dogs". His dog, with which these experiments were carried out, "had a very good nose". She easily followed his track even when it was overlaid by those of eleven other men and, on a perfectly calm day, instantly detected her master at a distance of 200 yards when he was standing in a deep dry ditch with only the top of his head above the level of the surrounding fields. Moreover, she likewise tracked him readily when his boots were previously soaked in oil of aniseed and, although she had difficulty in tracking him when he walked barefoot and seemed unable to do so when he walked, without boots, in new or previously worn socks, yet she found him without difficulty when the first part of his track was laid with boots although these were then removed and the walk continued, first in socks and then barefoot. On the other hand, she refused to track her master when he wore new boots, the boots of a stranger or even his own usual boots when the soles and sides were completely covered with a layer of brown paper and she readily tracked a stranger when he wore her master's boots, though she would not do so when he (the stranger) wore his own. In the experiments in which Romanes lined his boots with paper, the track was immediately picked up, however, at a point where "a small portion of the brown paper first became worn away at one of my heels... although the surface of shoe leather which touched the ground was not more than a few square millimetres".

Romanes' experiments were conducted in park-lands surrounding a house. Numerous trees, shrubberies, walls and the like provided cover behind which he could hide, unobserved, while he watched his dog during the whole course of each experiment. He took precautions, also, to ensure that she "should have to depend on her sense of smell alone". The very great interest which his investigations at once arouse makes it desirable to give some more of his own words. Here is his account of one of the most striking tests. "I collected all the men about the place and directed them to walk close behind one another in Indian file, each man taking care to place his feet in the footprints of his predecessor. In this procession, numbering twelve in all, I took the lead while the gamekeeper brought up the rear. When we had walked two hundred yards, I turned to the right followed by five of the men; and at the point where I turned to the right, the seventh man turned to the left, followed by all the remainder. The two parties thus formed, after having walked in opposite directions for a considerable distance, concealed themselves, and the bitch was put upon the common track of the whole party before the point of divergence. Following this common track with rapidity, she at first overshot the point of divergence; but quickly recovering it, without any hesitation chose the track which turned to the right. Yet in this case my footprints in the common track were overlaid by eleven others, and in the track to the right by five others. Moreover, as it was the gamekeeper who brought up the rear, and as in the absence of my trail she would always follow his, the fact of his scent being, so to speak, uppermost in the series, was shown in no way to disconcert the animal when following another familiar scent lowermost in the series."

Despite the interest which Romanes' work seems to have aroused, little or no progress in the investigation of tracking power was made during the following two decades. In Ghent, it is true, the world-renowned canine police force was established and some useful training methods (tracking on the leash is said to have originated in Belgium) were adopted and tests conducted in Belgium, England and Germany. About the year 1905, Brough, who was chairman of the Bloodhound Breeders Association, published a short article on the use of the bloodhound in tracking criminals. He pointed out that training should be begun when the dogs are very young, that every dog should be trained separately and that the difficulty of the tasks set should be gradually increased, great care being taken to prevent the animals from changing from one track to another. On good scenting days, he wrote, dogs may run as much as fifty yards to the leeward of the track. They should always be rewarded for successful performance. In the early stages of training, the tracks to be followed should be those of a person well known to the dogs. Later, strangers should lay the tracks which should eventually be appreciably more than two hours old. Brough insisted that bloodhounds could not be expected to follow a track which had been crossed, especially at the starting point, by many others and he ridiculed the idea that scents could be 'preserved'.

It would seem that Brough was an experienced breeder and trainer of bloodhounds but probably lacked scientific training. The opposite appears to be true of Zell to whose most valuable book reference has already been made. Although he reported some surprising successes of dogs in police service, Zell apparently consulted neither police dog leaders nor undertook tests but he was probably the first to study carefully many of the important factors which influence tracking power. Rejecting vibration theories of smell, he compared the olfactory power of dogs with that of other animals, contrasted the dog's 'world of olfaction' with the 'world of vision' and discussed the measures which should be taken at the site of a crime to prevent obliteration of tracks. Well trained dogs of suitable breeds, he maintained, could follow a human track several days old and he cleimed that dogs having good vision have poor olfactory power and *vice versa.* (This claim was disputed by several other writers including the author of the exhaustive work, *Der deutsche Schäferhund,* Von Stephanitz, who asserted that Alsatians, at least, have both good vision and good olfactory power. Von Stephanitz who was an adherent of the 'personal scent' theory, held that dogs can follow snow-covered tracks and tracks on ice. Tracks as much as ten to twelve hours or more old can be followed if conditions are favourable.)

Brough and Zell, apparently, did not know of Romanes' work but it was known to Richardson, the pioneer British trainer already mentioned who had vast experience with military and police dogs in Britain and abroad. He was the author of several books on dogs, perhaps the most relevant of which, entitled *War, Police and Watch Dogs* was published in 1910. Unfortunately, he was no scientist and, in referring to his own experiments, he omitted all important details. Knowing of the successes attributed to some dogs of the Berlin police and of the great decreases (35% or more) in crime in continental localities where police dogs were kept, he emphasized the great importance of putting the dog on the track as soon as possible after the discovery of the crime. He believed that bloodhounds were the best breed of tracker dogs, that they could follow barefoot and stockingfoot tracks more easily than tracks laid with shod feet, that cycle tracks could be followed if conditions were favourable, that water retains scent, that scent persists for hours (*not* for days) and that mounted horses can be tracked by dogs in virtue of horse scent (*not* the scent of the rider). The scent of blood, he observed, is very attractive to dogs.

Although Richardson's work seems to have made no great impression in British police circles, it certainly has not been forgotten.

Very different from Richardson's book is Friedo Schmidt's *Verbrecherspur und Polizeihund*, (Augsburg, 1910) which has already been mentioned above. This is a remarkable and valuable work in which great emphasis is laid on the necessity for the experimental control of hypotheses. The author made about two hundred experiments with the trained police dog Nixe von Stralsund (a bitch) which was tested on tracks ranging from 100 to 400 metres in length, no line being used. Believing that the dog's guide in following a human track and in identifying objects is the individual scent (Eigengeruch) of the track layer, he conducted experiments on the permeation of shoe soles by perspiration and examined the effects of various chemicals on the fatty acids which are, presumably, the chief sources of the odour. The guiding human scent, he pointed out, is derived chiefly from the perspiration of the feet but exists also on the hands and other parts of the body and is absorbed by the clothing. Odorous material is transmitted to the track having diffused through the foot coverings (boots, shoes, socks, stockings, if worn) being extracted by the moisture of the ground or other surface on which the track is laid. Schmidt remarked that criminals, for various reasons, and especially because they must often work rapidly and hard while under emotional stress, perspire more profusely than do innocent persons. Meteorological and other agencies, of course, obliterate scent either slowly or rapidly but he satisfied himself, by experiment, that it could not be masked by means of odours felt to be much stronger by man and he held that scent could be preserved for long periods (possibly months). He suggested, indeed, that tracker dogs could be assisted by spreading materials having such odours at places likely to be the scene of crimes. Attempts by criminals (or others) to prevent their production of scent or to destroy it, he affirmed, could not succeed. Schmidt believed, also, that dogs can readily follow tracks laid with rubber or wooden shoes or by the wheels of bicycles or carriages. Dogs could not be foiled, he maintained, by the use of new footwear or by using the footwear of another person because of the rapidity with which the individual scent penetrates leather and the ease with which it is transmitted by human hands and clothes to other objects. Concerning the dog's behaviour while tracking, he wrote that he could tell at a distance of several hundred metres whether or not the correct track had been found by watching the movements of the animal's (Alsatian's) tail.

Schmidt summarized his views in fifty conclusions which were subsequently published in English translation in the *Journal of the American Institute of Criminal Law and Criminology* in which, also, a favourable review of his book appeared. He himself published what is essentially a shortened form of his book in the *Jahresbericht des naturwissenschaftlichen Vereins zu Osnabrück*.

It would not now be contended that Schmidt's results were entirely trustworthy or that all his theories and conclusions were quite sound. In particular, the complete success attained in almost all his tests arouses suspicion and it is obvious that his tracks were much too short. Nevertheless, he was one of the first to study his subject profoundly and to conduct scientific tests. He was fully aware of the danger of over-estimating the reliability of canine evidence and be realized how much depended on the skill of the dog's trainers and leaders.

The books by Frey, Huyghebaert, Mathieu and Wagenknecht which quickly followed that of Schmidt, testified to the growing interest in police dogs and described, at considerable length, the various methods of training. But they did little, if anything, towards the solution of the olfactory problems concerned. Wagenknecht dealt with tracks up to 1500 metres long and four or five hours old. Huyghebaert, an author of long and wide experience in Belgium before the first world war, deplored the fact that French-speaking countries had fallen behind Germany in the study of canine 'olfactory memory' and remarked that, although for human beings, 'seeing is believing', the dog only believes what he smells! He devoted a section of his book to the question: How long does a track remain perceptible to the nose of a dog? and gave as his answer, more than forty-eight hours if circumstances are exceptionally favourable. It was his intention to write a book on the training of dogs but apparently he did not do so. He regarded homing by dogs as a consequence of the exercise of olfactory memory.

In Belgium, not long before Huyghebaert wrote, the Société Nationale pour l'Amélioration du Chien de Berger Belge, of which he was the secretary, had established standards of proficiency for tracker dogs. These required that dogs, not less than one year old, after sniffing at an object bearing the scent of the track layer, should successfully follow a track at least one hour old, remaining on it for at least thirty minutes and must indicate the track-layer amongst four persons, the track having been crossed at three places by that of a person other than the one to be followed.

The Société proposed to publish, in French and Flemish, a small brochure indicating the steps which should be taken at the scene of a crime when it was intended to employ tracker dogs, such a brochure having been produced in England at that time by Major Williams. It was suggested that the whole area should be isolated as far as possible and the strictest possible secrecy maintained concerning the crime. In this way, the assembling of crowds would be prevented and scent tracks would be protected from obliteration.

Meanwhile, in Germany, the use of tracker dogs by the police had given rise to much controversy and in 1913 and 1914, official tests, held in Berlin with dogs from all parts of the country, had such unsatisfactory results that the Prussian Government forbade the use of dogs in the criminal service. All the dogs tested were diverted by cross tracks if these were not too old, all tended to avoid change of direction and none could follow a track more than an hour or two old, pick out from amongst other objects those bearing scent, identify a person from an object bearing his scent or a scent-bearing object after sniffing at the person whose scent had been transmitted to it, or follow a fresh track in proximity to other fresh tracks. (According to Most, tests by the German Army authorities in 1925 and by the Munich police in 1927 yielded equally disappointing results.) Some tests were also made, at this time, at the Kaiser Wilhelm Institute for Brain Study.

These revelations and Most's contention that the human scent is *not* the factor which guides the tracking dog lead Craig, as we have seen, to maintain that the dog is completely discredited as a witness since it is thoroughly unreliable. Craig believed that both the successes and the failures of tracking dogs may often be traced to their power to observe minimal guiding movements (made by human beings). He also believed, however, that, in connection with canine tracking, there were great fields of interest awaiting investigation and that, as far as possible, tests should be conducted in the absence of human beings.

During the first world war, the study and training of tracker dogs other than for military purposes were presumably suspended but soon after the close of hostilities the work of the Dutch investigator Buytendijk, to which reference has already been made, was published in Holland. He first tested the very well trained dog Albert on a roughly quadrilateral course about 180 metres long on sandy ground, the track being laid in the absence of the dog and of the handler by a boy whose cap the dog was allowed to smell. Albert, running free, followed the track easily and successfully, appearing to allow for the strength of the wind. At first he kept his nose close to the ground but later held it higher. At one point he made a short cut across a corner in the track. In a second test, two persons laid a common track for some distance and then diverged at a right angle. Albert again tracked successfully after having smelled the hat of his quarry. He correctly selected from amongst three baskets containing odorous material that one which matched a fourth. Amongst the odorous materials were fatty acids including those occurring in human perspiration. After smelling at the hands of a person who threw pebbles amongst gravel for a distance of from ten to twenty metres, he invariably correctly retrieved the pebbles even when, before being thrown, they were handled by several other people. It seemed, moreover, that he could smell sodium chloride (1:10,000), quinine (1:10,000) and sulphuric acid (1 : 10,000,000). Inconclusive results were obtained with two other dogs (Bob and Nellie) which were offered food containing an odorous substance and a substance (quinine, for example) having a (for man) unpleasant taste. Tests were also conducted with boxes having swinging doors behind which food and an odorous substance were placed. When Bob and Nellie chose correctly, the door opened on pushing and the food could be taken. These two dogs readily distinguished plant odours such as eugenol and citral and one of them detected iodoform at a concentration of 1 : 4,000,000. They distinguished, although with difficulty, between the odours of nitrobenzene and benzaldehyde.

Buytendijk's report, unfortunately, lacks clarity in some places and the attractive book which he later wrote (see page 38) is also somewhat vague. He pointed out, however. that extreme caution must be employed in interpreting his results in view of the great complexity of the exciting stimuli and he emphasized the importance of the psychological factors involved in olfactory experiments on animals. A valuable review of his work was later published by Bierens de Haan, himself the author of a most interesting book entitled *Wat Gaat Er Om In Uw Hond?* Part of this book deals with canine tracking.

The adverse results of the official tests in Germany did not end the controversy which raged in that country nor did they interrupt the investigations which had been in progress. Thus it was that, in 1926, one of the few outstanding books on the training of tracker dogs was published. It was Blunk's *Die Ausbildung des Hundes zur Spurenreinheit.* The opinion that dogs cannot be rendered 'track faithful' (Spurenrein) and that their successes in tracking are due to accident or to influence by their handlers is, he asserted, quite erroneous and he claimed that he had proved it so. Bad results are due to training which is faulty, often grossly faulty. When training is good, the dog's natural habits are utilized immediately after it exhibits them and it learns to associate pleasant results with successful work. Like human beings, it avoids the unpleasant and seeks the pleasant and the trainer must take advantage of this. Actually, the dog is *naturally* track-faithful. What the trainer has to do is to teach it to be track-faithful *at the trainer's will,* to arouse the dog's interest in the work and to maintain it. This means, in practice, to cause the dog to pick up and hold to the desired track, avoiding all other tracks and indicating only those objects or persons carrying the particular scent which distinguishes the desired track. The attitude and skill of the trainer are, of course, of highest importance. The dog must solve the track problem itself and its handler, exercising extreme restraint, must never give signals to it in any way while it is on the track. No 'rummaging' should be allowed until the dog becomes track-faithful. There must be no retrieving by the dog (or the handler) of objects other than those bearing the scent being followed. In training, the master's track should not be used too often or when too fresh. Quite small objects (*e. g.*, pencils, pens, buttons, toothpicks) should be deposited on the track so as to make the dog sniff thoroughly. Complete strangers should always be used as helpers. The same track-layer should be used during each single day but the diversionary track-layers should be changed as frequently as possible. Blunk describes a training plan. In this, the person tracked must be and remain an "object of prey" to the dog. He also describes an 'indirect' training method in which the dog learns to select an object belonging to a suspect from amongst others belonging to strangers. This test is made in the absence of the suspect who cannot therefore exercise any influence on the dog or the test.

Blunk's work was only a part, although a particularly prominent part, of the research conducted in Germany in the years immediately preceding the 1914—18 war and in the inter-war period and it is worthy of note that, outside Germany, little or no scientific study of the subject took place during this time. German investigators of this period (Böttger, Brückner, Hansmann, the Menzels, Most, Schmid and others), in fact, have been responsible for almost all our knowledge of the mechanism of canine tracking and it seems convenient to discuss all their studies together, bearing in mind at the same time the careful and thorough studies of earlier workers and of Löhner, Neuhaus and others who carried out researches not of the field type. Amongst the German investigators, the most prominent by far was Konrad Most, at one time director of the Grünheide Training School for Police Dogs and later canine specialist in the German army. Most's studies of tracks *free from human scent* at Saarbrücken, Grünheide and elsewhere in 1909—14 and after the 1914—18 war caused him to expose the deficiencies of the methods of Friedo Schmidt and of the Berlin police who followed him and to institute rigorously controlled tests, asking himself two questions. First, what is a scent track for a dog and what odour or odours in it guide the animal? And, second, for how long does a human track (laid with shod feet) remain perceptible to the dog's olfactory sense? In one series of more than fifty tests with thirteen trained dogs, working under favourable conditions, he had observed that the track had been successfully followed on ten occasions only and that, on unfamiliar ground, the dogs were diverted by a single cross track! (In these tests, the dogs were not allowed first to take the scent of the track-layers who were, in fact, their masters.) Even less successful were some tests (in 1920—21) with army dogs which had been in training for more than a year — they could not even hold to the tracks of their masters. And some dogs passed without hesitation from a human track to a 'Fährtenbahn' (see below) track which they followed exactly as if it were a genuine human track. Most did not deny, of course, that dogs can easily distinguish the scent of one person from that of another. But he was not convinced that the guiding factor in the human track is that scent. After conducting tests with an artificial 'human' leg and foot, he constructed his 'Fährtenbahn', a large wheel to the periphery of which 'feet' made of wood, leather, glass and other materials could be fixed. This wheel, which had the weight of an average human being and was rendered completely free from human scent by heat (blow-lamp), was caused to roll along the ground, thus producing a track more or less closely resembling a human track but free from human scent. With similar precautions and under equally careful control, tests in which men hanging in chairs from a horizontal cable were drawn across a gap in a track and others in which men on stilts laid tracks were conducted. Several particularly well trained tracker dogs, the very best that could be selected, (Ito von Zirkelstein, Asta von Hohenlohe, and others) were tested off the leash in the presence of practiced observers who kept the whole experimental area constantly under supervision and made detailled records of performance and of prevailing conditions (weather, type of ground, and so on). Some tests in which tracks were produced with odorous materials allowed to fall from dropping bottles were also carried out and it was found that dogs could be trained to follow them. It was shown in these investigations that the dogs (not using optical clues) follow the tracks laid by the wheel and those laid by men on stilts and that they apparently perceive no track under the path of the men in the hanging chair. (They followed the ground tracks of these men to the point where the men climbed into the chair but went no further). Except when trained with special care, they always followed diversionary tracks (even purely optical tracks) when these were straight-line, or nearly straight-line continuations (*e. g.*, furrows, cycle tracks) of the track to be followed. On the other hand, certain specially chosen dogs (such as the Alsatians Ito von Zirklestein, Roland von der Baderstraße und some dogs belonging to the

Munich police) eventually became track-faithful on their master's tracks or even on the tracks of strangers (Ito von Zirklestein, for example, was engaged in nearly 1300 tests during a period of thirty-two months) whilst Blunk and the Menzels succeeded in training dogs to indicate 'identity of scent'. The Berlin police, however, did not attempt to achieve such identification nor did they recognize its possibility officially. Their dogs were usually able to distinguish tracks not less than three minutes different in age and they could hold to and follow tracks several kilometres long, rejecting all diversionary tracks provided that they never engaged in any other work. Some dogs were specially trained to search for living and dead persons, for hidden buried corpes and for shot game. (On one occasion, a dog successfully followed a track of drops of animal blood twenty-three hours old after heavy rain.) Most demonstrated with smoke clouds that air currents near the ground, which presumably carry scent, often change direction within short distances. He pointed out that the wheel tracks can carry the odour of crushed earth (and its animal and vegetable contents) or of crushed earth and vegetation only. The human track laid on the snow-covered ice of a lake with previously unworn shoes can carry, he said, the odour of the shoes only (leather). And the track laid on snow-covered ice by a barefoot person can carry the human species and individual odour only. Yet dogs, even when untrained, at once follow all these tracks. (The tracks on snow-covered ice were, however, fresh and short although invisible.) On the other hand, dogs which have become track-faithful on their masters' shod tracks, reject them if the masters' footwear has been worn for less than sixteen hours. But they detect their masters' scents even if the shoes are smeared with (to man) very strong smelling poultry droppings. If rubber shoes are worn, a dog can learn to follow a track laid with them and to ignore tracks laid with leather footwear.

Most had rightly begun his work by rejecting the false conceptions, uncritical assumptions and inadequately designed tests concerning the tracking powers of the dog (*e. g.*, that tracks as much as eight days old could be followed) which had been prevalent at the end of the XIXth century and are still current. (The very detailled 'protocols' published by the Deutscher Schäferhund-Verband in 1925 together with his essays entitled *„Vom Diensthund"* and *„Beiträge zur Frage der Verwendung von Hunden im Kriminaldienst"* showed that out of about 120 tests carried out with Airedales, Alsatians and Dobermanns, the best trackers in Germany, in 1913, 1914 and 1925 under strict scientific control in the presence of a great variety of experts, only two had successful results. One group of these tests was the series conducted by direction of the Prussian Minister of the Interior). The tasks set included: following fresh and old tracks, selecting wooden pegs bearing human scent, identifying persons who had imparted their scent to the pegs. Most also remarked, as several others have done, that olfactory acuity is not the only canine factor involved in successful tracking — adaptability, persistence, attention, loyalty, experience and very good memory are also required. He became convinced that, in general, the factor which guides the dog on the human track laid with shod feet is *not* and is not identical with the human species or individual scent and he regarded a period of about twenty-four hours as the maximum age of such a track which can be followed. The dog on the human shod track, he maintained, is guided by the *track odour* which may, for a time at least, include human species odour and individual odour that has penetrated the footwear. Track odour as usually produced by leather-shod feet consists chiefly of the scents emitted by crushed soil, plants, insects, bacteria and other organic matter together with those derived from leather, shoe-polish and any material which may adhere to the footwear. Many of these scents persist in the track, said Most, longer than does human scent. The crushing effect of the feet causes changes in moisture content, liberates volatile matter and initiates enzyme actions (fermentations) which may persist for several hours. Obviously, the track odour will vary in quality and strength from time to time and, according to the nature of the terrain and the meteorological conditions, from place to place. Dogs, of course, are accustomed to the changes which occur along the length of a track but, Most believed, they cannot distinguish the toe line from the heel line. On ground covered with bushes, trees and other similar vegetation, human scent may possibly be transmitted to the vegetation but, as a rule, such scent seems to play little or no part in guiding the tracking dog.

In his criticisms of the views and methods of some of those who subscribed to the individual odour theory of tracking, Most insisted that dogs should first be trained on one type of terrain only in the absence of tracks other than the one to be followed and, in the first instance, using their masters' tracks only. Their handlers should never know in advance the solution to the task set so that they cannot, even unwittingly, give clues to the dogs. Further, dogs should not be permitted to use visual and auditory clues. They should proceed from the easy to the more difficult task, not *vice versa*. Track-fidelity, even on the master's track, is far from easy to achieve. On the tracks of strangers, it is a task of very great difficulty.

In the final edition of his book (*Abrichtung des Hundes*, 1951. English translation, 1954), Most summarized his views on canine tracking based on the experience of a lifetime. The dog, he said, is not to be compared with an intelligent human pupil having human understanding and morals. Some dogs, however, can learn to persist on a single track, they can acquire track-fidelity. When so trained, they will, with a high degree of reliability, hold to human tracks several miles long if there are not too many other human tracks of the same age in the vicinity. The three-minute difference in age of tracks may secure persistence on the one track and a ten-minute difference will, as a rule, assure it so that large numbers of cross tracks will be ignored. Short distance tracking is possible on wet stone or asphalt but impossible on these materials when dry. In favourable weather, damp sand, even without vegetation, holds a scent track for periods up to twelve hours. The corresponding period for ground with grass and other vegetation is as much as twenty-four hours. Scent tracks are obliterated by heavy rain and, in perhaps three hours, by high wind and by sunshine. Although the dog may pick up the body-scent of its quarry when near it (on a track with sinuous curves, for example), the guiding factor is almost entirely *the scent of the track* in which the human species odour may sometimes play a part but the human individual odour, never. Hence, the dog does not indicate odour identities between track and quarry or between track and objects which may bear the individual scent of the quarry.

Most was well aware, of course, of the many occasions on which police dogs had successfully tracked criminals. He conceded that it is theoretically possible for a dog, in favourable circumstances, to learn to follow the human individual odour and he occasionally adduced evidence which would seem to support this view especially when considering the results obtained by the Menzels. But he affirmed that such results could be achieved by exceptional dogs only and, in company with some other German experts, he more or less completely rejected the individual odour theory.

When Cecil-Gurney visited Grünheide, the director, as we have seen, was Hansmann, a veterinary surgeon who had worked with the Berlin police and had played a very large part in the training of German police dogs. After Most, he was the most prominent of the German experts. He devoted much time to the study of canine olfactory physiology and psychology, pointing out the importance of such factors as association of 'ideas', the pleasure-pain principle associated with the perception of odours and the importance of memory in an animal in which smell is the predominant sense and that 'thinks with its nose' — an organ which is enormously more developed than is the human olfactory organ. Another factor which he stressed more than did other investigators was the importance of moisture on the dog's snout and on the track and of the solubility of those odorous substances which guide the tracking dog. Like other investigators, he commented on the great variations in olfactory acuity amongst individual dogs and on the harmful effect produced on the track by heavy rain.

Hansmann had early been impressed by the need to avoid misinterpretation of canine evidence. In 1928, he reported a case in which, because of serious errors made by several people, a child was falsely accused of theft on the basis of a dog's behaviour. Later, he recalled how the dog Harras (which subsequently passed into the possession of the Czar) had detected a murderer and how, also before 1914, a person convicted of murder on canine evidence was reprieved after Most as an expert witness, had aroused great interest by testifying against the dog. Still another case which he reported (in addition to the Seefeldt case — see page 73) was the following. A man went missing in June 1928. Firemen, police, school children and soldiers searched for him in vain. One month later, a police dog, brought to the search, found one of the man's shoes. The next day it dug up his corpse which was at that time completely hidden by heather from the view of those standing near.

In addition to case reports, Hansmann gave some illuminating and most interesting statistics. Thus, in the period 1926—31, ten dogs of the Berlin police were employed on 2833 cases. Of these, 1095 had to be excluded because of complete destruction of the tracks. In 566 (32.2%) of the remaining 1738 cases, the dogs were successful. In another period of three months, success was achieved in 18 out of 75 cases, a proportion of 24%. Errors by men, he remarked, are sometimes excused — by dogs, never!

Amongst the many questions which Hansmann asked himself, a significant one was the following: does the dog follow his master's individual scent, or that scent as modified by the dog's own sniffing over a long period, or, finally, its own scent which it has communicated in course of time to its master's shoes, clothing and so on.? This question, he believed, arose also if a person other than the master was used as quarry and was frequently sniffed at by the dog and he affirmed that dogs should *not* be trained on the tracks of their masters. An even more fundamental question, to which he replied in the affirmative (sometimes a hesitating, sometimes a decided affirmative) was: can the police tracker dog be employed in detective work with the assurance that it may give valuable evidence? He added, however, that this question must be examined scientifically and also in the light of practice without which the scientific approach is sterile. On their part, the practitioners must be modest and, when engaged as handlers in criminal investigation, they should always be in complete ignorance of the circumstances of the crime so that they cannot knowingly influence their dogs. If weather and ground conditions are favourable, the dogs will successfully follow tracks ten or more hours old. On the tracks, they should indicate (but, if possible, *not* pick up) objects such as stolen property, articles dropped or discarded by criminals and so on. On the basis of experience, Hansmann laid down a series of rules, regarding them, however, as subject to supplementation or even limitation. These rules were:

1. For the criminologist, the tracker dog is a *means* to an end, not an end in itself.

2. The dog is not a being with supernatural powers or a quadrupedal criminologist but a technical instrument amongst others. It cannot 'search' for a burgler or murderer.

3. Only rarely does the dog lead from the scene of the crime to the criminal (Hansmann himself recorded one such case). Hence, training should be for the usual not the exceptional case.

4. Halting or barking at the supposed criminal should be forbidden. This rule should be observed from the beginning of training. If the dog accidentally halts or barks at a person during tracking (*e. g.*, when following a criminal in flight from a crime just committed), no notice should be taken of the fact — every dog will do this occasionally especially to shy or anxious people — and all present should at once be told to ignore it. (In Hansmann's view, Most's practice of teaching dogs to regard the track-layer — real criminal or mere substitute — as an enemy and to bark at or even bite him was quite wrong. It might even be dangerous and might cause the animal not to be track-faithful. Most, Hansmann continued, stopped his enquiries 'at the most important point'.)

5. Searching for a criminal is the concern of human beings and there is no point in having it done by the dog.

6. The police dog's duty is solely to find and follow a human track, not perceptible to man, the start and finish of which are of concern to the police or criminologist as the track of a criminal or of a harmless person, not to search for a person or identify him. Pursuing this duty, the dog indicates direction, follows the track partly or completely, finds objects, leads to buildings, and so on. Whether and to what extent objects found on the track or in buildings or elsewhere and whether persons living in such buildings or temporarily present in them have any connection with crime or criminal is for the criminologist to determine.

7. For technical reasons, the tracking dog must be kept on the line (except when following a criminal in flight from a crime just committed).

It is worthy of note, said Hansmann, that the training of tracker dogs is chiefly negative, *i. e.*, they must learn to hold on each occasion to one track, ignoring all other stimuli however attractive to them. They must 'specialise' early and learn to keep the nose close to the ground. But, he contended, it is rarely, if ever, that the human species odour or individual odour, which may not persist for long in the track, is the guiding factor for the dog and, indeed, a trained track-faithful dog will follow a six-hour-old human track four to six kilometres long, *ignoring diversionary tracks laid ten minutes before by the same person.* What guides the dog is the specific *track odour* present in every human track which distinguishes tracks from each other. Thus the tracks of human beings differ not because of differences in their individual human scents but because every track has its own specific odour and because, in so many cases, the tracks differ also in age, Moreover, it is possible for a track which has become so dry that the dog cannot follow it to be so renewed by moisture (and not too high temperature) that it can then be followed. Well trained dogs, he affirmed, can even follow tracks on snow, on hard-frozen ground and, for short distances, on paved surfaces. But there are no scent-faithful (fährtenrein) dogs and there never has been one.

Summing up his views shortly before the second world war, Hansmann maintained that dogs can be bred and trained to be track-faithful (or track-certain) so that they will hold to fresh and old human tracks ignoring fresh and old diversionary tracks. Some dogs, he held, can learn to do this with almost complete certainty. (His longest experimental track to be successfully followed measured seven kilometres and was seventen hours old.) In actual practice, *i. e.*, in tracking leading to the recovery of stolen property, to the elucidation of criminal acts and to the identification of criminals, the difficulties encountered are much greater than in laboratory or field experiments and the proportion of successes will, probably, not exceed 30%. (The 'prophylactic' effect of success by police dogs is very pronounced — crime soon diminishes!) It is possible that the human individual odour may play some part in guiding the dog and, indeed, it has been found that, where dogs have 'lost interest' and abandoned tracking, their interest may be aroused again if they find on the track an object bearing the scent of the track-layer. (Other objects, however, will sometimes produce the same result!) Perhaps it will be possible to show by experiment what part, if any, individual odour may play. In the meantime, the tracker dog should *not* be given, for sniffing, an object believed to bear the scent of the quarry but should be set on the track at a point where it is practically certain that the quarry has tarried. The tracking power of dogs is, in any case, one of the most interesting of all biological phenomena and this remains true no matter what view is taken of the various theories of tracking so far propounded.

During the years 1949—1955, Hansmann again reviewed his work at Grünheide and elsewhere since its inception in 1902 and pointed out that, in the post-war period (after 1945) almost all the mistakes previously made were being repeated and almost all the knowledge gained (in Germany) during the years 1911—1933 ignored. His observations of tens of thousands of dogs had convinced him that dogs having the right temperament and 'very goods noses' can successfully follow a human track as much as thirty-six hours old. But it must not be forgotten that such tracking is very hard physical and 'psychical' work. For the tracker dog, shoes and other clothing are part of the human being and, in practice, it is almost invariably the clothed and shod person with which it has to deal. It has *not,* he maintained, been proved that human individual odour penetrates shoes and clothing but it *has* been shown, and can easily be confirmed by experiment, that most dogs do not know their own masters when they are naked; that no dog recognizes his naked master, just bathed, by odour only unless it has systematically learned to do so; that no dog, though long shown to be trustworthy on the track, recognizes its master by the odour of his bare feet; that dogs 'recognize' a stranger wearing their masters' clothes and regard him as their master until they first hear the stranger's voice; that dogs immediately recognize the clothing of their masters and that of other known persons; that dogs treat their masters as strangers and even, in some cases, attack them blindly if they wear the worn clothing of strangers. The dogs are then confused when they hear the master's voice. It can also be shown that dogs protect a stranger wearing their masters' clothes from attack by their naked masters. These facts make it difficult to believe either in the individual odour which penetrates the footwear and clothing and remains for hours on the track or in the 'anxiety' sweat (or, even, the 'death' sweat) said to be produced by anxious criminals and others. In 1925, Hansmann had stated in an official memorandum to the Prussian Minister of the Interior that adequately trained tracker dogs can follow human tracks hours old without changing to fresher and older diversionary tracks and that criminal experience had confirmed this. But that every human being transmits to his track his special individual scent which the dog can smell and follow so that if it encounters the individual anywhere it will halt him and bark at him — this is simply assumption! The decisive factor for the dog's nose is *not* the scent of the track-layer *but* the scent of the track which is constantly changing and may even be stronger after ten hours than after ten minutes. In any case, the dog should never be allowed to halt and bark at a suspected person. The dog is not a criminal investigator and it is not its function to find the criminal. It is rather one of the 'instruments' to be used by the investigator.

The memorandum was accepted and, in consequence, rules for the employment of tracker dogs and for the prevention of their misuse drawn up. After the memorandum had been adopted at an official conference in Munich in 1931, police dogs achieved great and growing success throughout Germany.

In common with several other investigators, Hansmann considered in detail the nature and composition of footprints and tracks and the innumerable biochemical changes which continually occur in them. These changes, presumably, destroy an individual human odour but produce a specific track odour perceptible to the dog's nose and distinguishable by it from other odours. Thus, in a test with three old tracks (laid at 10 p. m., 1 a. m. and 4 a. m., respectively) and three fresh tracks (laid at 7.0, 7.30 and 7.45 a. m., respectively), all laid by the same person, all crossing each other at many points, and all having many angles, a dog, set on one of the old tracks, followed it correctly, ignoring all the others. Further, objects laid on the track which the dog was following by another than the track-layer were indicated just as if they bore the scent of the track-layer. Hence, objects and track do *not* necessarly have a common scent component and a dog *cannot* show whether or not objects found on or near the track bear the scent of the track-layer. (As we shall see, the tests and successes of the Menzels were *without* a track.) In tests with Most's tracking cross, with two or more track-layers, Hansmann's dogs almost invariably changed to diversionary tracks at the point of crossing. He observed this on hundreds of occasions but only when the tracks were on similar types of ground. When they were on different kinds of ground, the dogs were much more successful especially if the track-layers differed appreciably in weight. Success increased to 100% when tracks were one hour or more old and differed in age by not less than from three to five minutes.

Even if human species and/or individual odour does pass to the track, it is absolutely certain that the track odour is different from the human odour and it would not be wise, criminologically, to train the dog solely to this. For, if dogs were trained to follow a single 'odour characteristic' (or signal) they would become 'mechanized' which is worse than becoming 'anthropomorphized'. It is true that dogs have been known to follow a track from the scene of a crime to the location of the criminal apparently on the basis of a single particular odour *(e. g.,* that of leather). But this is quite exceptional.

Those who have lived with dogs and watched them closely for long periods, wrote Hansmann, know that their intelligence differs from that of human beings and that it cannot be properly or fully tested in laboratories or in the field. Moreover, there still remains an immense amount to be learned about their many-sided behaviour, capabilities, and skill.

One of Hansmann's colleagues who was 'in at the beginning' in Germany was Paul Böttger. As early as 1926, he published a booklet giving instructions for the training of tracker dogs and describing in detail more than twenty tests, all but one successful, on dogs of the Berlin police. In these tests, tracks as much as fifteen hours old were followed. In a later publication, he reviewed the history of police dogs in Germany, dealing with some of the work of Most, Hansmann and others, discussing methods of training and reporting results. He recalled that Most had contended in a memorandum to the authorities dated 15 January 1913 that it had not so far been found possible, through training, to cause a dog by letting it take a particular scent, to devote its attention to that scent, to follow the corresponding track and to bark at the person who laid it or to retrieve an object bearing the scent concerned. Likewise, it had not been possible by training to render a dog track-faithful *(i. e.,* to acquire track-fidelity) so that it would follow the track of a particular person without changing to other human tracks. Also, it had been shown that tracks six hours old or more cause very great difficulties for the dog in most cases. Böttger also recalled that, in the Prussian official tests at Grünheide in 1913—14, it was required that the dogs should remain on a stranger's track even when it was continued by others of the same age, should follow old or very old tracks (4—17½ hrs) and should select from a collection of wooden clothes-pegs that one which bore the scent of a particular person after taking the scent at that person.

As has already been stated, *all the dogs tested failed.* (Böttger gave diagrams of the tracks and of the courses followed by the dogs.) It was evident that the dogs paid no attention at all to the tracks. Later tests (by Most) in 1925 and 1927 were no more successful. In 1927—29, tracking tests were conducted under the same conditions with highly trained tracker dogs of the Berlin police. This time, most of the dogs failed *but not all.* Success increased when the dogs were given the opportunity to become familiar with the odour of the track. This was done by causing the track-layer to trample for three minutes a surface of half a square metre at the start of the track where the dog was induced to sniff thoroughly. Tracks more than two hours old, however, were very seldom successfully followed. All this showed the necessity for *correct training* and in a test instituted on May 15, 1928 the Alsatian Ito von Zirkelstein, rejecting all continuing tracks of the same age laid by strangers, followed her master's track successfully. However, Böttger was unsuccessful in getting the dog, after taking the scent six metres from the starting point, to select and follow the correct track (her master's) from among six hundred-metre 'identical' tracks with five metres between each pair but, of course, laid by six people. Other similar tests were also not successful and Böttger held that detecting a track-layer from his track or the track from the track-layer (stranger in both cases) had not then been definitely proved possible.

After Hansmann had been made director at Grünheide (1926) and had improved the standard of tracking considerably, he established a police dog training and investigation section at Ravensteinmühle near Berlin. Böttger, who became director of this section, recorded certain conclusions which he had reached concerning training. Small earth-coloured objects should be laid at long intervals on the track to stimulate the desire to track. The track should have many bends and angles. During training, the dog should work on the *long leash and harness* and no 'free searching' should be allowed. Moreover, the dog should not be permitted to take the scent in the air *(i. e.,* blown by the wind) as happens often in hunting. If this is permitted, the odours *in the track* tend to be neglected.

At Ravensteinmühle, the aim was to train dogs to follow old tracks several kilometres long ignoring younger and older diversionary tracks and ignoring also other disturbing factors *(e. g.,* men, road traffic). Much success was achieved and errors were eliminated. From the beginning of its training, Böttger maintained, the dog must work on ground where there is traffic so that it becomes accustomed to cross-tracks and other disturbing factors. For each success *(e. g.,* when one of the small objects is found and indicated), it must be given great praise. These objects not only mark out the track but also cause the dog to keep his nose down and to hold to the track. As a result of this training, dogs learned to hold to a pre-determined human track many hours old and crossed by older and fresher tracks. In July 1926, fifteen-hour old tracks were followed. (Even before this date, Moll had shown that dogs could follow perfectly stranger's tracks eleven hours old and severely complicated with fresher and older cross tracks; and Blunk, in 1926 had trained a dog to indicate odour identity of a person with an object, no track intervening.) In 1928, Böttger continued, the Menzels exhibited two boxer dogs which could invariably select specified objects by scent from a large number of others. These results prove, he said, that when no track odours are involved, it is only the human individual odour which serves for identification.

Dealing with the effect of type of ground, Böttger pointed out that Berlin police tracker dogs of highest skill could not follow scent on stone or asphalt surfaces or on hot dry sand free from vegetation. On moist sand without vegetation, they followed tracks up to six hours old. On land with vegetation in moist air and moderate winds, tracks up to twenty-four hours old or more were followed. On grass, in hot dry air with strong sunshine, he said, scent is destroyed in three hours. Meteorological conditions, of course, greatly influence tracking in various ways. On snow, scent lasts longer than on ground without it. But if it falls on a track and covers it to a depth of more than four centimetres, it interferes with or prevents tracking. Heavy frost is not unfavourable but heavy rain is. Slight rain, however, is favourable. Dried-out tracks are reactivated by moisture.

Tracks differing in age by less than three minutes cannot be distinguished from each other by dogs but their power to distinguish increases with age-difference above this period.

Böttger believed that the odour followed on the track is a combination of human scent with the scent of the track. Artificial odours, he held, should *not* be used in training. As regards weather effects and the use of dog's evidence, he expressed agreement with Friedo Schmidt. It is not yet known, he affirmed, what the greatest age of a track which can be followed is. A dog trained to follow stranger's tracks ignores his master's. So does a dog trained on its master's but later put on stranger's tracks. At the scence of a crime, those who propose to use tracker dogs must obviously act quickly to prevent trampling by sight-seers and others. For the dog, the duty is to pick up and follow the track, *not* to identify *(e. g.,* by barking at the track-layer) the criminal. In order to illustrate his views, Böttger quoted from Berlin police records as follows. During 1929—32, the dogs of this police force helped in the search when 4611 persons were arrested. Of these, 1087 were 'wanted'. Fifteen persons were prevented from fleeing. He also gave some of the reasons why actual practice is often much better than training. These were: 1. the criminal often remains for a considerable time at the scene; 2. he travels lonely routes as far as possible; 3. he avoids other people and conceals himself when necessary behind natural cover; 4. the weight of his booty makes his footprints pronounced and his scent also.

Böttger recorded further that the Prussian Minister of the Interior had called an All-Germany Conference on the Use of Police Dogs in 1927 and recalled that the findings of this conference had resulted in the formulation of official rules in 1931. Since 1906, he pointed out, the Berlin police had never falsely accused anyone on canine evidence or caused any false suspicion to fall on anyone. On the contrary, many innocent people, suspected or arrested wrongly for other reasons, had been freed from guilt by the evidence of dogs, the true criminals being found and imprisoned.

Amongst the many German investigators of canine olfactory power, the married couple Dr Rudolf and Dr Rudolfine Menzel occupied a place of special importance. It is probable, indeed, that they have surpassed all other workers in this field and that, in some respects at least, their dogs (and dogs trained by their exact physiological and psychological methods) have achieved greater success than have any others. It was their view that man, over-estimating his place in the animal kingdom, is too anthropocentric to judge animal intelligence objectively. Intelligence is not necessarily only human and just as animals have abilities which man has not so they have an intelligence of their own. Further, man has no appropriate units or standards of measurement for animal intelligence. It is true, of course, that many, perhaps most, human aims can never be appreciated by the dog. Of all the members of the animal kingdom, however, there is only one which really loves man and makes man's interests his own even when man treats him harshly — the dog. As a result of long symbiosis with man, the dog can learn to abstract from a sphere foreign to its own necessities. In general, an animal's power of abstraction is strictly limited to its own sphere of interest. But the dog can be taught to let its interests partially coincide with — or at least overlap — those of man in such a way that it acquires at least a dim conception of its trainer's aim. It is this ability which must be taken as the only basis of training if success is to be attained, the process of utilizing the dog's olfactory powers in indicating 'identity of odours' being as much psychological as physiological. Man utilizes the dog's olfactory ability because his own is inadequate, because only thus can he appreciate indirectly a world which, more or less, is forever closed to him — the world of odours. In taking advantage of the dog's natural habits (retrieving, tracking, and so on) the trainer must begin with the very young animal (this is essential), maintaining sustained attention throughout working life, never allowing keenness to flag, developing a strong understanding (rapport) between trainer and pupil, communicating to the pupil the idea of 'human individual odour' and causing the pupil to come to regard its work as an interesting sport. Thus, those habits which are servicable to man are channelled and encouraged whilst others are (not too severely) suppressed. In particular, the dog must learn to hold fast to the guiding factor, rigorously ignoring diversions *however attractive*. Not every dog can acquire the skill necessary for tracking or for other olfactory work. The chosen animals must possess drive in searching, pleasure in retrieving, spirit, hardness and willingness to be guided. And only first-class trainers will achieve much success. It is because of the inadequacy of other trainers that so many misunderstandings have arisen concerning the powers of dogs and that many people have rejected their employment as trackers. At the best, the careful and intensive training necessary is very slow and takes a long time (years). Moreover, it must be remembered that there are two aspects of canine olfactory power, namely, ability to indicate scent-identity and ability to track. These abilities do not depend solely on olfactory acuity, other factors such as instinct, heredity, breeding and selection being involved.

For the interested scientist not less than for the criminologist and police officer, the primary question concerning the use of tracker dogs is: Does the tracking dog follow the human individual scent and can it distinguish several human individual scents from each other and concentrate on one only. From this question many others arise. Perhaps the most important of these are:

1. has each human being his or her own distinctive individual scent?

2. can the dog perceive these scents, distinguish each of them from the others and remember them?

3. can man, making use of the dog's power, so influence the dog that it appreciates which individual human scent is, on each occasion, of interest?

4. can man arouse in the dog the willingness to make man's interest its interest also?

5. is the dog able so to concentrate on a given task that it will not be diverted by other interests some of which are much closer to its nature?

The Menzels gave unqualified affirmative answers to the first two of these questions , regarding their view as based on proof. Their answer to the third was yes, for those who understand and take the necessary trouble. To the fourth they replied: yes, but only for those who are in closest intimacy with their dogs and to the fifth, also yes, if the dog's interest is adequate and persistent and if it enjoys the task as a human being might. It is liable, of course, to be diverted by stronger interests but can learn to ignore these for a time. And they added, in support of their answers, that they had possessed dogs able to do all these things. Many of those who had given negative answers, the Menzels held, had made the fundamental mistakes of supposing that the dog in *naturally* interested, especially at the beginning of training, in concentrating its attention on human individual scent, of failing to allow the dog sufficient time to take the scent at man or object and of changing the quarry too often. Except in the case of the dog's master or some other well known person, however, this interest does not naturally exist. The individual scent of the master is central to the dog's pleasure and interest and hence it is on this that training should begin. The dog should learn first to follow its master's track (in the absence, at first, of diversionary tracks) and to search for and retrieve pieces of clothing and other objects bearing its master's scent. When it has thoroughly mastered all the tasks based on its master's scent, but not before, it should then leave them completely and learn to recognize the scent of another person, preferably a familiar person, becoming able, eventually, to recognize him in any clothing and all objects bearing his scent, ignoring all secondary odours such as those of cloth, leather, shoe-polish, tobacco and wood.

It has been proved, the Menzels affirmed, that the trained dog when tracking a human being follows the individual scent which remains on the track and, as Most showed, is generally mixed with the odours of soil, of crushed vegetation and other organic matter and of shoe-leather and shoe-polish. However much the impression of this mixed odour changes with time, temperature, humidity and physiological and psychological condition in track-layer or dog, it retains its characteristic feature just as a human face does whether angry, or laughing, or weeping. It has also been proved that, if the track is long enough (usually about 800 metres or more), the dog can identify the track-layer and objects bearing his scent, selecting him and them from amongst other persons and objects. Further, the dog can be trained to distinguish the human individual odour in the absence of a track after thorough taking of the scent and, if there is a track of sufficient length, to distinguish that scent *without* taking scent at the person concerned. In other words, dogs can become 'scent-faithful' (witterungsrein) and also, although this is peculiarly difficult and has probably seldom been achieved under certain conditions only, 'track-faithful' (spurenrein).

The Menzels believed that the length of time during which human individual odour persists in the track is not known. They were also unable to state the minimum time required by the dog to 'fix' a human individual odour in its mind, how long it can retain it in memory and how much time it requires to adjust to another individual odour. They pointed out however that a dog which has learned to follow one specific scent cannot quickly or readily learn to follow another. On the other hand, they maintained that despite the 'decomposition' odours which many organic materials yield on keeping it is possible for a dog, if specially trained, to recognize a human individual odour *(e. g.*, that of a suspect or criminal) on an object *(e. g.*, article of clothing) suitably preserved in a glass container in a dry and cool place for as long as six months. But they added that, if possible, a control object should be taken at the same time and from the same place (part of body) bearing the scent of a person known to be innocent and that a test with some other 'odour preserve' should be made a day or two before the actual criminal investigation. The reports of tests and police cases which they gave provide considerable support for their view.

In their book "*Die Verwertung der Riechfähigkeit des Hundes im Dienste der Menschheit*" published in 1930, the Menzels give full details of their methods of training in various types of scent recognition and identification: object from man (with and without a track), man from track, man from object, object from track, object from object, track from track, track from object. The insisted that objects to be scented should be in contact with the scent-giving source for not less than ten minutes and they remarked that when a particular scent-bearing object is to be selected from others also bearing scent, the distances between the objects should be sufficiently great to prevent scent being 'carried over' from one object to another. When scent is taken from a human being, the armpit is the preferred place.

Obviously, some of the above types of recognition and identification are appreciably more difficult than others and, in fact only a very few, even amongst the most highly trained dogs, can sometimes succeed in these. Scent-differences between tracks, they said, are usually smaller than those between objects and are correspondingly more difficult to detect. On the other hand, it is not peculiarly difficult to train dogs to complete success in identifying a person from an object bearing his scent.

It was the Menzels contention that since, during walking, the human foot (more usually, the human footwear) is in contact with the ground for less than one second at each step, the scent in a single footprint must be subliminal. But they were convinced that, on the track, summation of scent stimuli (subliminal stimuli included) occurs and hence that the tracking dog can familiarize itself with the scent provided the track is long enough (600—1000 metres for the not-too-old track of a stranger). The stimuli, too, need not occur in a continuous series.

The obstacles to successful scent-recognition and scent-identification, so frequently discussed by others, were naturally not forgotten by the Menzels. They held that the skilled dog would be able satisfactorily to relate human regional odours to individual odour, to overcome the difficulties caused by various types of terrain and, of course, to ignore diversionary tracks since, in all situations, their sole guide must be the human individual odour. Tracking in actual practice, they affirmed, should not be so difficult as in tests where conditions could not but be unnatural. They rejected the suggestion, made by some investigators that the canine odour is a source of error. The well trained dog ignores its own 'possessions' (*e. g.*, collar) and, when following a specific track, the tracks of other dogs as well as that of its own master and of other dog owners despite the fact that the clothes of dog owners will probably carry some canine scent. They also stressed the point that on all occasions of success, praise and reward should be on a most generous scale. Finally, with so many others interested in the tracking of criminals, they urged that as even the most successful dog is only one criminological 'instrument' amongst others, the greatest caution and restraint must be exercised in evaluating its findings.

After the second world war, Dr. Rudolfine Menzel studied the duration of persistence of olfactory tracks on various kinds of soil (apparently in Israel) and showed that in the detection of buried explosive mines, dogs are more efficient than are instruments because they smell, not the mines, but the 'wound' made in the ground when the mines are buried.

In the controversy concerning the olfactory powers of tracker dogs which raged in Germany before 1914 and especially between the world wars, many people both inside and outside the ranks of the police and the military took part. One of the results of this was that, in the period 1927—36, several valuable papers on the subject appeared in the pages of the *Kriminalistische Monatshefte.* Moll, the author of the first of these publications, had previously exposed the alleged abilities of the horse 'Clever Hans'. He made the important point that ability to evaluate the methods and results of tracking practice and tests does not depend on experience in handling or training dogs. It is enough to know the principles and conditions of exact science and to be able to detect the sources of error liable to occur when experiments are designed and conducted. In particular, he stressed that conscious or unconscious guiding of the dog by its master or handler must be excluded by keeping them in ignorance of the course of the track. Using five dogs of the Berlin police, he arranged more than fifty tests, two of which he described in detail. Most of the tracks included bends with diversionary tracks continuing straight parts. One track was eleven hours old. Since the dogs successfully followed the tracks 37 times out of fifty-four (*i. e.*, 68% success) and since they also retrieved small objects some of which had been in contact with the track-layer for 20—30 seconds only (others were in contact for long periods) he concluded that although no dog is absolutely track-faithful, some dogs, if trained for one type or work only and having the right temperament, can learn to be so through training only and not through intelligence, except when circumstances are particularly unfavourable. Such circumstances, Moll held, may frequently occur in practice so that success at a level of 68% is not to be expected. He also held that the human individual odour (which is probably not identical with the odour of the feet or of some other part of the body) plays some part in guiding the tracking dog and in enabling it to retrieve objects bearing human scent. In order to determine the effect of artificially applied odorous substances, he had parts of tracks laid by persons whose boots carried clove or other essential oil. Sometimes the dogs seemed to be quite uninfluenced by the odours of these substances but he did not collect enough evidence to enable him to express an opinion concerning their effects on tracking.

Some of Moll's views were shared by Krack whose communication followed two years later. He claimed that the attempts of the sceptics to prove that the dog is incapable of following the human individual odour lacked scientific basis and he rejected as unproven the view that the tracking dog follows the odour of crushed vegetation on the ground and that carnivores react far more to animal than to plant odours. Further, he maintained that the results of tests made with the track-laying wheel are by no means unequivocal. On the other hand, he admitted that because dogs have a very labile psyche, the existence of track fidelity had not been proved. Since the dog can have no interest in a criminal's track as such, *we* require it to develop interest for *our* purposes, taking it for granted that there is no canine conflict between *wish* and *duty*. But whether or not this conflict occurs, wrote Krack, the attitude and behaviour of the dog's master is of immense importance and, if the bond between him and his dog is very strong, astonishing results may be achieved.

Seeger, the next contributor, began by warning against expecting too much of the dog and failing to appreciate the differences between tests in laboratory or field and actual practice. He regarded it as proved that the specially trained dog recognises the human species odour and can pick up and correctly follow the human track, indicating or retrieving objects bearing human scent. Limitations on this ability due to the nature of the terrain and the weather should be remembered, he added, and he listed the rules concerning the use of dogs in tracking criminals issued some years before 1929 by the Württemberg police, these rules being regarded as indicating the pre-conditions for success. They were: 1. reasonably undisturbed tracks or objects from which the scent can be taken must be available and not more than six hours must have elapsed since the estimated time of the crime; 2. the site of the crime, at which the scent is taken and from which the track is followed must not be exposed to much traffic; 3. since the crime was committed there must have been no very adverse weather (rain, snow, fog). In special circumstances, the first rule could be ignored. In Seeger's opinion, it was doubtful whether the dog can indicate scent-identity ("Witterungsübereinstimmung") and it usually does not follow the whole length of a criminal's track especially as its quarry may have used transport (even aircraft!). Reporting a case (criminal damage to fruit trees at night) where a dog followed the long tracks of subsequently convicted criminals after 16 and 19 hours, he pointed out that nevertheless the dog's evidence was misused since the most that the animal could show was that the person concerned had been at the site of the crime about the time when it was committed — before, during or after was not revealed by the dog.

The author of the next communication in this series was Reiss. Having had very wide experience of innumerable cases and having scarcely ever noted during thirty years a conviction based solely on the evidence of a dog, he asked why should the use of dogs as trackers of criminals simply be abandoned. Most's claim that tracker dogs readily change from one human track to another and make other similar mistakes could, if substantiated, be regarded as the 'funeral oration' of the police dog — to train him would be to waste time, to use him would be foolish. Reiss maintained, however, that a thoroughly trained dog kept constantly in practice — there are very few such, he believed — under the care of a highly skilled handler who is in complete harmony with his beast *can* hold to a human track without being diverted by fresher or stronger tracks human or animal, provided certain climatic, meteorological and other conditions are fulfilled. (If they are not, the handler should refuse to allow his dog to be used.) There must be, of course, a foot-track which can, with reasonable probability be held to be that of the criminal and, unless this is the only track unfamiliar to the dog, he must be set on it as he cannot himself select it. It may help the dog if he can take the scent from an object which very probably belonged to the criminal or bears his scent and can then pick up the track nearby. Only in exceptional cases, however, will this happen especially as the object may bear other scents in addition. Moreover, objects of glass, metal, stone and similar materials do not long retain scent. Tracking on asphalt, concrete, paving and the like is possible for short distances only. The maximum age at which a track can be followed varies considerably. Reiss knew of a favourable case in which it was twenty-one hours. He believed that the guiding factor for the dog was the human individual odour and maintained that the odours of earth, crushed vegetation and shoe-leather had no significance. And he remarked, concerning the dog's natural inability to appreciate human guilt that, if an innocent and a guilty person were present at the site of a crime at the same time, the dog might as readily indicate the robbed as the robber.

The last communication of the series was from Böttger some of whose work has already been considered. He insisted that the training of police dogs should be specialized, trackers being trained differently from searchers which are used *(e. g.,* to detect corpses, lost persons, objects) when no tracks are found. It has not been proved, he repeated, that human odours guide the dog — tracks laid by barefoot human beings are *less* easy to follow than are those laid with shod feet. From a knowledge of his track-odour a dog *cannot* pick out a stranger from among others nor can it distinguish a stranger's track from equal-age tracks of other strangers. Similarly, no dog can select from human tracks of the same age, that of the person whose scent has been taken at an object bearing the scent. With Most, Böttger came to hold the view that it is the odour of the track which is significant and which the dog should be trained to follow. The smallest track age-difference which a dog can distinguish, he maintained, is three minutes. After referring again to the important effects of weather conditions and to the elucidation of the Seefeldt murders, he recorded cases of the successful tracking to their homes of some poultry thieves and laid down the following principles of tracking.

The dog can be trained to be track-faithful so that it ignores fresher and older human tracks and follows old and long human tracks.

The dog's task is to follow the track of a particular person using his track-odour as guide. The dog must be set on the track at a point where the quarry has spent some time moving about.

The dog may work out part of the track only but can indicate objects, signs and other factors useful to the detective. It is the detective's task to evaluate the findings and to determine whether a retrieved object belongs to the quarry or is retrieved merely because it bears human scent. It is a serious error to incriminate a person merely because the dog barks at him.

At the scene of a crime and on the track, all indications should be protected to prevent interference or damage. If the dog halts because it reaches a point where it can no longer follow the track (on a concrete road with heavy traffic, for example) it should be urged on. If this fails, the detective must continue without the dog's help.

In an earlier publication, Böttger had affirmed that the training of tracker dogs should begin at the early age of four months and should become intensive at about one year. As training proceeds, the difficulties of the track (shape, number of diversions, *etc.*) should be gradually increased. He preferred to use bitches (Airedales, Alsatians) but regarded males as equally efficient. They should, he said, always work in special harmess on a long leash (but free from influence by the handler) keeping the nose to the ground, avoiding over-strain and resting when necessary. The training area should have many tracks. It should not be paved or have crowds, many onlookers or much traffic. In order to assist the training dog, the quarry should move round on a surface area of half-a-square metre for a time at the starting point and should then set out dragging the feet for about ten paces. The dog must not be allowed to bark at or attack the quarry. Objects to be left on or near the track should be scented by carrying in the trouser pockets. Tracks should be fresh or old and may be as long as about 8500 paces. Complete records containing all relevant details should be kept. The age of the tracks may be as much as fifteen hours.

Of all the tests which he then recorded (about twenty-five in number), all but one were successful.

The tracker dog's task, Böttger stated, is to disclose what man's senses cannot. Its evidence must be weighed exactly as is all other evidence. He had never encountered a judge who convicted on dog's evidence alone.

In a still later publication (1937), Böttger reiterated his assertion that the human individual odour is usually not the factor which guides the dog tracking a human being and referred to tests which he had carried out as early as 1906 with dogs of the Berlin police. If the individual odour were the guiding factor, the training dog should be able: 1. to pick out and follow the track of a stranger from amongst those of other strangers and to identify the track-layer amongst the others. 2. while tracking, to search for objects bearing the scent concerned when these objects are amongst similar but unscented (or differently scented) objects placed at the same time, and 3. to take the scent from a strange object *(e. g.,* jacket, stocking, trousers), find the track to which it is related amongst tracks of the same age, follow the correct track and identify the person concerned amongst a number of other strangers — all this under strictly defined conditions. However, by laying simultaneously two or more zig-zag tracks or curving tracks which cross each other frequently, it can be shown that dogs cannot do these things although, if the tracks are adjacent but do not touch or cross, many dogs succeed. Böttger conceded that dogs trained by the methods of Blunk and the Menzels can learn to recognize the individual odour of a stranger and to select an object belonging to him from amongst a number of other unfamiliar objects *provided the person concerned is present.* They cannot, however, identify the person from his track. On the other hand, a well trained dog can follow a strange track over fresher and older diversionary tracks if it first allowed to become familiar with the track-odour and, as would be expected, the ease of following is increased as the difference in age of the tracks increases. Tracks of heavy persons, too, are more easily followed than are those of light.

It is in actual practice, rather than in experimental tests, Böttger stated, that dogs really learn. He regarded it as very imporant, also, that they should be allowed to follow their natural instincts as far as is compatible with the requirements of their work and he gave some illustrations of such work. In four cases of house-breaking near Berlin, four dogs with four different handlers working at intervals of weeks all tracked to the same spot where they then lost the track in heavy traffic. The criminal concerned, however, was eventually found in that area and confessed. Again, in 1908, a Berlin police dog followed the track (length three kilometres, age about six hours) of a hanged woman, revealing footmarks and a bicycle. The owner of the bicycle was found, confessed and was sentenced whilst another person, at first accused of the crime and arrested, was now released as not guilty.

Finally, Böttger discussed the training of dogs in neighbourhood searching rather than in tracking and reported that, during a short period, Berlin police dogs found fourteen out of sixteen missing persons, men, women and children.

In 1938, a particularly valuable contribution to the question "Is the individual odour the guiding factor for the dog in recognizing his master?" was published by Niemand. During a period of about two years, he tested 103 dogs taking very great care to exclude sources of error. He found that the dogs paid more attention to a stranger wearing their master's clothes than to their masters wearing the clothes of strangers and that they were apparently unable to recognize their naked masters by scent although ability to do this increased with length of association of master with dog. He also conducted tests with naked men who had ingested sulphur and with others who had been sprayed with lavender and examined the effect of washing (no soap used) on human scent. The conclusion drawn from these experiments was that the odour of the naked human body is not identical with that emitted by clothing which has been worn although the two odours apparently have or acquire a common component which a dog can learn to regard as the specific odour of the individual concerned and hence, eventually, to recognize him both when naked and when clothed.

Still another German investigator who published a useful contribution in 1938 was Major Belleville of the Berlin police. He provided some illuminating statistics. These showed that, during a period of five years, in 1458 tests with 4404 diversions 4114 were correctly rejected, nine dogs having been concerned in the 290 cases of diversion. The proportion of successes varied from 86 to 97%, being 91% in the case of barefoot tracks. Even when weather was unfavourable, the proportion was 83%. Since some dogs, at least, can distinguish human tracks of the same age in addition to those of different ages, it seems certain, Belleville maintained, that human tracks contain the species and also the regional and individual odours as distinguishing components. One dog, set with the wind behind him on a 200 metre track at the end of which identical pairs of objects made of cloth, glass, iron, marble or wood were deposited was 90% successful in selecting the glass object bearing the scent of a human hand, the other object of the pair having been sterilized. In the case of articles of cloth, the proportion of successes was only 45%. The degree of success in determining the direction taken by a track-layer was only 47% when dogs were led to the mid-point of a track along a line at right angles to that track. Since one hundred tests of this kind were made, it seems possible that the direction is often found only by accident. Belleville suggested that when tests are made by Most's 'cross tracks' method, the track-layers should take paces of equal length and walk at the same rate and he also made the following recommendations: track-layers should wear their usual but not new footwear; the dog's handler during a test should not know the solution of the problem set; owners, judges and others taking part in tests should be so placed as to be unable to influence the dogs during test; test grounds should have the usual human and animal tracks of various ages (Böttger had pointed out that without such tracks they are useless for training); dogs should be encouraged to work with enthusiasm but not forced to work if disinclined.

Discussing various types of test (barefoot, with rubber footwear, with track-layers who change shoes at some point on the track) he recorded that when two men laid one and the same track carrying each other alternately on the back, one dog ceased to follow the track at the point where its quarry was taken on the other man's back. Not unexpectedly, he found that human beings with strong scent are more easily followed than are other human beings and they also cause more errors when they lay diversionary tracks.

As to the maximum age of tracks which can be followed, Belleville mentioned twelve hours but added that success is sometimes achieved on older tracks. Human tracks overlaid with other human tracks and parallel pairs of tracks of the same age close together can be followed he said, seldom, never or for short distances only.

The last German investigator whose work will be considered is Bastain Schmid, a professor in Munich who disputed Most's views. He maintained that dogs, although not track-faithful by nature can learn to become so or, at least nearly so on the tracks of their masters and also become so on the tracks of strangers although under certain conditions only. We know, he wrote, "from tens of thousands of experiments with hundreds of dogs that the following of a track by a dog is a much more complicated process than the mere following of the individual odour of the quarry". The first dog to become track-faithful on its master's track was the bitch Ito von Zirkelstein which could distinguish his (Herr Schubert's) track from that of his wife. When he wore new boots, never before worn, Ito showed uncertainty. At the time Schmid wrote, the only dogs known to be track-faithful on stranger's tracks on any kind of ground were those of the Munich police (*e. g.*, Tessi and Vagabund). Vagabund, although friendly with a companion bitch, ignored her track when it was a diversionary track. Schmid regarded it as proved that the dog can select individual animal odour from a complex odour. Moreover, it can detect by vision stationary objects at distances up to 585 metres and moving objects at distances exceeding 810 metres, its hearing is much more acute than is human hearing and, sometimes at least, it can even distinguish the sex of some animals by scent. (When the wind is strong at ground level, he stated, body scent may be carried as much as sixty metres aside.) Tests on hard frozen ground covered with snow (depth up to 20 cm, temperature sometimes as low as —23° C.) showed him that the snow extracts odorous material from footwear (and also from the paws of dogs) and so facilitates tracking. Asphalt and dry stone, on the other hand, prevent tracking. (In order to prevent dogs from using *visual* clues in snow, they were sometimes blindfolded. This retarded but did not otherwise affect them.) In one series of tests in which one or more diversionary tracks were laid, success was achieved in 85% of the cases. Schmid also conducted twenty-two tests on the tracks of six ridden horses (male and female) chiefly on hard frozen snow-covered ground (no hoof-marks visible to man). In twenty-one of these, the dogs were successful but all his tests with horses on open ground failed. He concluded that they, like dogs and human beings emit individual odours (not mixed with the odours of footwear or clothing!). He recorded also that he could himself detect hoof-odour and 'metallic' odour in horses' hoofprints.

Peculiarly interesting were the results of a few tests which he made on tracks of 100—200 metres with two barefoot men. One of these, with both feet unwashed, hopped 73 metres on one foot and continued 78 metres on the other. The tracking dog followed the whole course of the track indicating the point of change. When the test was repeated after one foot had been thoroughly washed with sand in running water, the results was the same but the dog worked more slowly. In later tests, the point of change was sometimes indicated. When tests with the two men alternately carrying each other (like those conducted by Belleville) were made, the dog sometimes followed the whole track and sometimes stopped at the point of change. Repetition of the test after the feet had been washed revealed that the dog stopped at the point of change.

Schmid suggested that tests of these kinds should be repeated and extended. His conclusion was that dogs could solve all the problems if appropriately and adequately trained. As far as human individual odour is concerned, he held that the material responsible penetrates the leather of new unworn footwear within 22 hours. He pointed out, moreover, that adult human skin, which can never be freed from excretions for more than a short period, loses as much as 14 grammes of 'scales' (dead cells, *etc.)* daily and that other animals likewise secrete or excrete odorous material of this kind regularly. Again, since even the *human* nose can detect the difference between a patch of ground which has been footprinted and one which has not, it is understandable and, in fact, it has been proved that dogs can be trained to follow specific human and animal individual odours in the mixed odours of the track, ignoring those of other human beings and animals. Experiments carried out by Schmid showed that puppies aged forty days have a well developed sense of smell. They ignore unscented cloth but eagerly select pieces bearing various odours (human scent, meat). His dogs correctly tracked other dogs for distances up to 300 metres on snow and on open ground where there were cross-tracks of other dogs and of birds, deer, foxes and hares. And he watched a wolf, with nose to the ground, follow the track of a strange dog only half-a-metre in front.

Amongst the odorous substances which attract lively canine interest, the lower fatty acids occupy a prominent position and, as Neuhaus, Moulton and others have found, very low or even minute concentrations of these compounds can be detected by dogs. In Japan, during recent years, field experiments in which these acids were employed have been conducted by Uchida who had observed that his dogs preferred scents in the following increasing order: the scent of a stranger, the odour of butyric acid, the scent of the dog's trainer, the odour of caproic or caprylic acid. Having noted that butyric acid 'arrests' tracking dogs and that they have the habit of rolling on evil-smelling (to man) carrion on which they rub their necks and shoulders, he arranged for strangers to lay tracks 1000 metres long and marked each of these at two points with fatty acids obtained from putrified earthworms. After fifteen minutes, his dogs (36 Alsatians, male and female, which had passed the Japanese police dog tests) sometimes successfully followed the tracks but they often stopped at the marked points, licked them, urinated, rolled on the ground and abandoned the tracking. When those which had abandoned the tracking were again set on the track, they usually followed it to the end but repeated the actions at the marked points.

In a second series of tests on snow, pairs of track-layers (the dogs' masters and strangers) wore rubber boots smeared in some cases with fatty acids. The two 200-metre tracks A-0-B and D-0-C formed a tracking cross (Mosts's cross) and at the common central point 0, one of the track-layers discarded his smeared boots, replacing them with clean ones. Those dogs which were not well trained abandoned their tracking when 'diverted' by butyric acid. Dogs well trained in tracking strangers exhibited the preferences given above. Some of them successfully followed the correct track of master and stranger, not being diverted by the butryic acid odour. However, this odour, for inadequately trained dogs, and the odour of the higher acids for all dogs prevented any or correct tracking by all the dogs.

Reference has already been made to the experimental work of Krushinskii and his co-workers (p. 32). In addition to the laboratory tests which they undertook, they conducted some tracking tests, tracks 300—600 metres long being laid. The dogs (more than twenty were used) were of both sexes and aged from one to seven years. Age and sex, however, did not affect acuity. These workers later tested three groups of tracker dogs (133 dogs in all). They found that continued systematic training in the recognition of one odour *(e. g.,* human scent) did not increase acuity to other odours *(e. g.,* odour of meat) and they drew a distinction between acuity alone and tracking power, although these characteristics are usually fairly closely correlated. For measuring tracking power, they used a four-point scale and they rejected as unsuitable for tracking, all dogs which could not detect the meat odour through seven layers of fabric or whose tracking power did not reach point four on their scale.

Almost all reports of tracking by police dogs refer to European conditions. Accordingly, the accounts of criminal procedure in Egypt (from 1936 onwards) given by Sir Thomas Russell Pasha in 1949 are of particular interest. In a country where fear of the results of giving evidence was very widespread, excellent service was rendered in the tracking of brigands and murderers and others by the dogs (Alsatians) which, in addition, often struck terror into guilty persons, causing them to confess. Russell gave details of cases in which dogs successfully followed human and donkey tracks several days old. His reports are of unusual value but it is impossible to be sure how far they demonstrate canine olfactory capabilities.

In 1875, Galton suggested that it would be interesting to know whether dogs could distinguish twins by scent. Eighty years later, some experiments bearing on this question were undertaken in England by Kalmus. He used nine dogs which varied greatly in intelligence, perseverence and degree of training. Six of them were Alsatians (four male police dogs and two show bitches), one was a Doberman Pincher (a male police dog), one an old male Labrador (also a police dog) and one a German pointer. His scent-givers were seventeen men, nine women and five children, all but one, a male African, being European. The dogs' first task was to single out a handkerchief scented by a particular person or to find the track of a particular person amongst from four to ten other human tracks. Handkerchiefs were scented in armpits and the dogs took the scent at the hands of the scent-givers. At first, all but one handkerchief were unscented. A few of the dogs proved very good in the handkerchief-retrieving tests. In the tracking tests, the track-layers walked side by side for 100 yards in a field and one then dropped a handkerchief, glove, cap or similar object. Then, after a further 25 yards, they fanned out and, on reaching a distant hedge or row of trees, they hid themselves. The dogs were then brought to the starting point of the track which they followed, picked up the object and, if successful, continued on the track of the person who had scented the object. The police dogs were quite reliable in this test.

In a second series of tests, the dogs had to select the track of a person who dropped an object from amongst five diversionary tracks all laid by one other person. One well trained police dog made very few mistakes in this test when the two people were not identical twins. Four pairs of identical twins (three female, one male) then took part in both retrieving and tracking tests. Sometimes the handkerchiefs of both twins were used, sometimes the handkerchief of one only. The evidence of one dog, Chloë, showed that the scents of the twins were treated as identical. In tracking tests, however, done as before, (one correct, five divergent tracks), one dog usually succeeded in distinguishing between identical twins. Kalmus concluded that well trained dogs can, apparently, distinguish the scents of identical twins in some cases, but, when the scent of one twin is offered in its absence, the dog readily follows that of the other twin. The dog also recognizes an object bearing one human regional odour after taking the scent at another region.

If the tests were repeated and extended with appreciably larger numbers of dogs and people, very valuable results might be obtained.

Canine Societies

In Germany (as in Belgium and, no doubt, elsewhere) canine tracking tests were conducted or discussed not only by individuals but also by various societies, the most important of which were, probably, the *Schäferhund Verein* and the *Gesellschaft für Hundeforschung*. Thus, in 1928, the

Schäferhund Verein published reports of tests on irregular tracks aged up to four hours having lengths up to 1100 paces. In most of these tests, the dogs were successful and in another series of trials with dogs trained by the Menzels, there were no failures. These dogs invariably selected correctly objects bearing the scent of a particular person after taking the scent at that person's boots, coat or socks. They did this even when a handkerchief bearing Frau Menzel's scent was amongst the other objects and they refused to select an object if the one bearing the 'correct' scent was not present. Moreover, they picked out from a group of recumbent persons that one whose scent they had previously taken.

The reports of the Gesellschaft für Hundeforschung, published in 1932, dealt with a questionnaire sent to its members in 1931 and with its first general conference in the same year. Relevant points raised in the questionnaire were:

1. What influence have air pressure, air movement, temperature and atmospheric moisture on an odour-track?

2. In what relationship are the 'odour molecules' of the human regional odours to the total individual odour of a person? Are there quantitative differences only or also qualitative?

3. Have children less pronounced individual scents because of their sexual, immaturity?

4. What is the basis for the so-called 'theory of anxiety-perspiration'?

5. Are there odour differences in the individual odours of male and female persons, sick and healthy persons, alcoholics and temperate persons?

6. What are the 'minimum perceptible' of odours?

7. Does the dog recognize only those objects belonging to the scent-giver which he has touched with the hand and not those worn on the body but not touched by the hand?

8. Do dogs which hold the head up while tracking follow the same odour as do those which keep the nose close to the ground?

9. How is the dog's power usually to follow the toe-line of a track to be explained?

10. What is the shortest time-difference so far determined between two tracks of different ages which a dog can distinguish?

11. Does the dog follow a barefoot track more easily than a track laid with shod feet?

12. What effect has the following of long tracks (fatigue?) on the ability of dogs to receive impressions?

Participants in the four-day, first general meeting of the Society inspected the army canine school at Kummersdorf, the police canine school at Grünheide (director: Hansmann) and the railway canine school. Lectures were given by Konrad Most and others, reports on tracking by skilled dogs were presented and some of the dogs (Rolf von Schülerheim, Asta aus der

Wolfsgartenstrasse, Ito von Zirkelstein, Edda, Lygia beth Satan, Ulysses ben Satan, Ala al Din, and others) demonstrated their ability to follow tracks and to distinguish scent, so providing a partial answer to Most's important question: How can the dog be caused to concentrate its attention on that odorous constituent which is significant for the purpose involved?

As on many other occasions, Most stressed his view concerning the common factor usually present in all tracks on land, namely, the odour of crushed plants and pressed earth. In a demonstration of the 'suspended man' test on the first day, a dog followed the foot track of a man to the point where he sat in the suspended chair but did not follow the route of the chair. Another dog followed the foot track and continued along the prolonging track laid by odour-free porcelain 'feet' on a rolling wheel. Both these dogs had never before undergone such tests and had never been on the test ground before. A third dog, which had learned to follow its master's fresh track correctly under easy conditions in 93% of tests made in one year and had later succeeded in scoring 60% in following a stranger's fresh track (also under easy conditions) was exhibited. A score of 90% was reached by a fourth dog in following tracks laid by a man wearing rubber shoes, diversionary tracks being laid by men wearing leather footwear. Some tests were made on tracks produced by dropping odorous liquids. In these tests, the dogs were successful even when conditions were difficult. A complicated track, 1300 metres long, laid with ox blood was followed with complete success by a one-and-a-half year old bitch which had previously been successful on ox-blood tracks as much as 31 hours old and 2.5 kilometres long.

On the second day of the meeting, the bitch Edda took only a few minutes more than one hour to follow a difficult track 3150 metres long, sixteen hours old and crossed by diversionary tracks, picking up seven objects belonging to the track on the way. (She received some help at a point where workmen had damaged or destroyed part of the track.)

One dog, trained by the Menzels, correctly followed a track one-and-a-half hours old and selected first amongst four vests and then amongst four handkerchiefs the vest and handkerchief bearing the scent of the track layer. Another of the Menzel's dogs, after sniffing at the armpit and breast of some person, selected from amongst four objects the one bearing the scent of that person. This dog was completely successful in all the fourteen tests made and also, even, in a fifteenth in which one of the four objects bore the presumably most familiar scent of the trainer, this object, of course, being ignored. Apparently, however, this same dog did not recognize articles which had long been worn by the person with whose scent it was familiar after these articles had been sealed in glass containers for two weeks but it familiarized itself, by 62 seconds of sniffiing, with the scent of a stranger and identified objects bearing this scent.

Still another of the Menzel's dogs which had been trained throughout its life to respond to the odour of an aromatic ether, was completely successful in selecting an object bearing this odour in the presence of three other objects which were sometimes scent-free, sometimes scented with chemicals (menthol, turpentine, bergamot oil), and sometimes scented with human scent. In tests of this kind, however, this dog was known to fail occasionally.

It cannot be denied that, at least under certain conditions, dogs can perform most astonishing tracking feats, following both known persons and strangers with a high degree of success. Thus, at the World Congress of Kynologists at Frankfurt a. M. in 1935, two dogs belonging to the Munich police repeatedly tracked a stranger with perfect success, ignoring in the process cross and continuing tracks of the same age. Similarly, dogs of the Berlin police, ignoring all other tracks, were able to follow the half-hour-old tracks of their masters under most difficult conditions with almost complete success. They could also distinguish tracks not less than three minutes different in age and, to quote a somewhat different example from criminal procedure (not tracking), they detected the corpses of two murdered boys in a very thick wood when all other methods of search had failed (the Seefeldt case). Some details of this case are worth recalling.

In the latter part of February 1935, two schoolboys in Mecklenburg, Germany, disappeared without trace. A large-scale search was at once begun in a plantation containing almost impenetrable thickets in which it seemed possible that a certain Adolf Seefeldt, who was under suspicion of having murdered ten other boys, might have concealed the corpses. Although a large number of police and other persons took part in the search and although eight not specially trained police dogs and their handlers searched the area thoroughly for about two months, no bodies were found. On May 31, however, search dogs of the Berlin police were brought to the scene. Between 8 a.m. when they began to search and noon, they also were unsuccessful, but soon after noon they showed by scratching the ground at a certain point that they had detected something. This proved to be the corpse of one of the missing boys buried at a depth of twelve inches. The burial had been conducted with extreme cunning, an intact sheet of moss having been cut out with a sharp instrument. After the body had been interred, all surplus earth had been completely removed and the moss had been replaced exactly in its original position. New growth had then completely eradicated all signs of a disturbed surface.

Despite very unfavourable weather (great heat and absence of wind) the search for the second boy continued until June 7 when it was suspended for a week because of the Whitsun holiday. A few days after the holiday period, however, heavy rain fell, improving conditions for scent perception and when the rain ceased on June 20 the dogs soon found the body of the second boy buried in the same cunning manner at a depth of sixteen inches. Thus did dogs show, beyond any reasonable doubt that murder had been committed, also locating the corpses which might otherwise have remained hidden for many years, if not for ever.

Again, Schmid's experiments seem to show that dogs can successfully track horses (on hard frozen ground but perhaps not on soft ground) and other dogs of both sexes (for short distances at least) in open country traversed by deer, other dogs, foxes, other horses and men. This suggests that they can distinguish both 'species' and 'individual' odour in dog and horse. And it is certain that any odour which may serve as guide in these cases is not complicated by any odour of footwear or clothing.

Such achievements, of which the tracking of a stranger is probably by far the most difficult, require long and tedious training of the dogs concerned and the necessary skill is not often attained.

How then are the many failures and contraditions to be explained? Why, for example, are the tracks of stilts and artificial feet followed whilst the suspended man is not? Perhaps a partial answer is as follows. The dog employs not only his nose but all his powers in tracking and if his successes and failures are to be understood, the olfactory power must not be regarded in isolation. In so far as this power is concerned, the guiding factors seem to be the scent of the track and the body-scent (species, regional, individual) of the quarry. The scent of the track, like that of the body which is presumably present normally also on the track, is complex. As Most pointed out, it is emitted by plants, insects, bacteria and so forth crushed by the feet of the quarry. This crushing causes liberation of volatile matter and changes in moisture content and initiates enzyme actions (fermentations) which may last for some hours. As has already been said, additional odours usually present in the human track are those of leather and shoe-polish. The scent of the track will vary greatly from place to place and from time to time.

It seems probable that the scent left by man in his footprints is a mixture in which the individual odour is only one, although perhaps the most important, component. The scent of the human body is affected by that of the clothing, by the use of detergents, cosmetics and the like, by the kinds of food and drink consumed, and by many other factors of greater or less importance. In view of all this, little credence can be given to reports that tracks as much as, for example, a week old have been successfully followed. (And it may be doubted legitimately whether Odesseus in his beggar's clothes was recognized by his old dog Argus after an absence of nineteen years!)

It is not surprising, either, that the untrained dog is an untrustworthy or completely incapable tracker, and, on the other hand, that if a dog is successfully to perform the most difficult feats it must be specially trained in one type of duty only. It may be taught to take the scent from a man and then to identify objects to which his scent has been transmitted. Or it may be given the more difficult task of identifying the man from the object or from the track.

It can be understood, too, why the dog will follow a track such as that laid by the 'feet' on the wheel in which there is no human scent and why it can follow a scent from which all but one (or not much more than one) of the factors usually present have been eliminated.

Some remarks may now fittingly be made about olfaction in other animals. But first reference must be made to some additional investigations which concern the dog.

Rather more than ten years ago, Dutch workers found that head nystagmus could be induced in dogs, fixed on a board, by rotating in front of them a turn-table on which rabbits were placed at intervals. When the sense of smell of the dogs was suppressed — but not when their vision was obstructed — the nystagmus disappeared. This nystagmus seems to be olfactory in nature but, since odour is not directional, the phenomenon is difficult to understand.

Other significant facts of a different kind were later revealed, also in Holland, by Kloek. Having found that steroid sex hormones have odours, for man, resembling those of human and animal excretions, he tested their effect on dogs of the canine school of the Hague police. For this purpose, he used aluminium rods which had been held for five minutes in the hands by men, women, boys and girls and jars containing material bearing the odours of the hormones. On the basis of the frequencies with which the dogs selected rods he concluded that steroids are excreted through the human skin and that dogs can distinguish men from women, boys from girls, and so on, by smell. It would seem, however, that the precautions against error taken by Kloek were much less stringent than those taken in many of the tests conducted in Germany many years earlier.

Perhaps the most recent attempt (1962) to determine the olfactory threshold in dogs has been that of Becker and his coworkers. The dogs were accomodated in boxes, arranged in a circle. At the centre of the circle, a powerful fan dispersed the odour of two substances classed as resinous according to Henning. The substances, which differed significantly in quality of odour (as judged by man) and had no biological significance for dogs, were dissolved in alcohol and one drop of the solution, at various concentrations, was allowed to evaporate on a glass slide. Several dogs were tested but most of the experiments were done with a spitz and a toy collie. They were given complete freedom of choice and were rewarded with meat when successful. The highest concentrations offered to the dogs were not perceptible to human beings. The thresholds found were 4.14×10^{-4} and 7.69×10^{-4} mg per drop.

A rather different matter was examined, as long ago as 1920, by Szymanski. Wishing to know whether men also could follow an odour trail, he laid trails with ichthyol on sheets of paper which he then attached to frames holding wire netting. He found that ten men, by sniffing at the netting, could follow the trails.

Amongst the other animals (not birds, fish or insects) which have been studied from the olfactory point of view, rats take a prominent place[10]. They have been tested in recent years by Foster, Gordon, Eayrs, Moulton, Tribe and, especially, Le Magnen. These investigators have found that rats easily learn to distinguish between odours which are similar to man and apparently do not confuse odours which are confused by human beings; that they distinguish between stereo-isomerides *(e. g., d-* and *l*-camphor); that they do not seem to share human views on pleasantness (methyl salicylate) and unpleasantness (butyric acid) in odours and that their power to detect lower aliphatic alcohols and aliphatic acetates in homologous series increases with increase in the length of the carbon chain of these compounds. Strangely enough, it was also found that, for the rat, smell is *not* a very important factor in the choice of food. Moulton thought that this is possibly so because smell is more a biophysical than a biochemical sense.

Experiments on normal monkeys by Masserman and his colleagues indicate that these animals are apparently not much affected by odours. They seem to prefer fruity odours but show no aversion to the stinks of indole or even skunk. Neurotic monkeys, on the other hand, react very vigorously to odours after learning to associate them with the neurosis-producing agent.

About thirty years ago, the olfactory powers of guinea-pigs were examined by Matthes. In experiments with nitrobenzene and bromostyrene he found that their olfactory acuity was about 1000 times as great as that of human beings and that it increases very greatly and very rapidly with training. They could detect ten drops of a solution of one part of these compounds in ten million parts of solvent when mixed with 80 g. of oatmeal porridge.

Cattle, hedgehogs, frogs and mice have also been studied, although not at all extensively, during the years since the second world war. Perhaps the most interesting point which has come to light as a result of some of these studies is the effect on breeding of the "social" odours which affect communities of mice and, perhaps, of birds and other animals also.

In concluding this chapter, we may mention, first, that in 1936, Oldfield published a very informative account of the scent glands of the

[10] In 1907, some rather crude olfactory tests were carried out in Germany with the notorious horse "Clever Hans". His likes and dislikes were examined with ammonia, ethyl acetate, iodoform, nitrobenzene, orange, peppermint, phenol, turpentine, vanillin and a few other substances.

alligator, beaver, civet, crocodile, fox, gecko, goat, hippopotamus, lizard, musk deer, musk rat, peccary and pig; and, second, that a 1962 U.S. Patent (No. 3,046,192) describes a scent for attracting wild game. This scent consists of musk, urine of fox, deer or other animal, peppermint and aniseed on a porous material which is heated to body temperature and placed in a suitable position (*e. g.*, hung on the branch of a tree). Finally, it should be recorded that the British-American project mentioned earlier in this chapter is perhaps being continued. Becker and his colleagues have now found, by testing two dogs with glass slides lightly imprinted with human odour from the finger, that, when the slides were kept indoors, the animals could easily detect the odour over periods of up to three weeks and that they were fairly successful even after six weeks. When the slides were kept outdoors, detection was easy up to 24 hours and fair up to 96 hours.

Experiments on tracking by dogs are still being continued at the present time. In Canada, the police of Vancouver, in collaboration with Hamilton Wright, have given enthusiastic reports of their value in locating and controlling criminals and, in the U.S.A., Fuller and his co-workers have examined various breeds that were induced to search for hidden persons and objects. But the tracks followed in these tests were very short and several important factors that influence tracking were not considered. Tests of tracking and scent recognition have also been undertaken recently by the Japanese workers Kusaka and Mori. Children, adults and horses, their report states, acted as quarry and the tracks were on asphalt, concrete, grass and sand. The persons tracked wore various kinds of footwear (*e. g.*, canvas shoes, plastic or rubber boots, wooden sandals) with and without socks or walked barefoot. It concluded that "personal scent will not exist in scent left by footgear". However, since the English version of the report is obscure, it is not possible from it properly to evaluate the conclusions reached especially as no reference is made to temperature, humidity or strength or direction of wind.

Perhaps it may be appropriate to bring this section to an end by quoting a remark by Adrian. "I do not think", he says, "there is any direct evidence to show that the absolute sensitivity of the organ to various pure substances is greater in a dog than in a man", and perhaps it may be added that, according to Gruch, rats, whether wild or tame, are also not appreciably, if at all, more sensitive than are human beings.

However that may be, it seems very probable that the mathematical analysis of systems of olfactory communication amongst moths and ants employed by Bossert and Wilson (1963) could be applied, with equal success, to the behaviour of dogs and, no doubt, to that of birds and fish also.

Wheresoever the Carcase Is

For wheresoever the carcase is, there will the eagles be gathered together.
Matthew, xxiv, v. 28

I am forced to the conclusion that the animals [ducks, parakeets, pigeons, siskins] I have used ... do not possess the faculty of smelling. *Walter*

Les oiseaux sont donc doués du sens de l'odorat à un degré tout au moins égal à celui du chien, pour ne citer qu'un example connu de tout le monde, et c'est une bien grosse erreur qui subsiste encore dans la littérature scientifique, que celle qui représente ces animaux, pourtant pourvus d'un appareil olfactif très complet, comme incapables de découvrir leur nourriture autrement que par la vue. *Raspail*

The olfactory powers of dogs are not disputed — not so those of birds[1]. When Audubon "contradicted all former opinions on the subject" in Edinburgh in 1826, he did so with uneasy feelings expecting to encounter vigorous opposition from those who then accepted the general belief that some birds, at least, have an acute or even a very acute sense of smell. In the biblical view, it seems to be implied that it is the smell of the carcase that attracts the eagles. And although the weight of evidence based on observation and experiment indicates the doubtful character of this opinion, it has by no means been abandoned, even today.

Audubon's experiments were made with vultures (turkey buzzards, *Cathartes aura)* in the Southern States of North America. Having noticed on many occasions that these birds repeatedly pass over without notice objects calculated to stimulate their voracious appetites because these objects are unseen; and also that they apparently failed to detect his presence by smell, he decided to conduct experiments designed to show whether or not they, in fact, possessed an acute sense of smell. The results of these investigations caused him to give up the notion that the vulture discovers its prey at an immense distance by means of its sense of smell. He found that a vulture which again and again attacked a stuffed deer, being for long unable to realize that it would yield no flesh, nevertheless instantly detected a living snake apparently having no odour and scarcely as thick as a man's finger at a distance of several hundreds of yards. He concealed a dead hog in a ravine twenty feet deep, covered the carcase with cane, and left it for two days at the height of summer. Although he found the smell of the putrifying flesh insufferable at a distance of thirty yards, many vultures flew over the ravine in all directions in search of food but none discovered the carcase. And he observed that two caged vultures which customarily exhibited great excitement when they saw him approach with or without

[1] Some ornithologists have affirmed that sitting birds are scentless, *i. e.*, free from the odour which they naturally possess at other times.

food, remained perfectly quiet when he came near without being seen or heard and when fresh or putrid meat was brought near them but out of their sight.

Audubon had set out to explode a generally entertained opinion and it is not surprising to learn that the opposition which he anticipated actually developed.

One of his severest critics was Waterton who, having been "a wanderer in South America", felt himself to be in a position to publish some rather sarcastic comments. He was able, indeed, to reveal various contradictions and deficiencies in Audubon's reports and, since he believed that vultures never kill the animals on which they feed and that they never eat food which is not putrid, he insisted that only by smell could they distinguish putrid from fresh meat and dead from living (but not moving) animals. Why, he asked, did Audubon fail to state whether he was up or down wind from the vultures which he watched? How could he see the small snake at such a distance? Was he certain that his stuffed deer was quite odourless? How was it that his keen-sighted vultures failed to see and join the dogs feeding on the putrid hog carcase which they could not smell? (It could not be because of fear of the dogs, since, according to Audubon himself, dogs *and* vultures have been seen feeding simultaneously on the same carcase.) Could he explain how high-soaring vultures can distinguish satiated members of their species *(e. g.,* in trees) from others who are hungry and are waiting an opportunity to feed or how the vulture high in air knows whether the prey of another on the ground is big enough to make a descent worth while especially if the other flew down in error and found little or nothing? Did Audubon really believe that vultures alight near the putrifying carcases of alligators but fail to feed on them because (apparently) they do not know that parts of the carcases are soft enough to be penetrated? Finally, asked Waterton, how did Audubon succeed in coming so near to the vultures which he watched without being seen by birds, which, as he himself points out, are very readily frightened off when human beings approach? Waterton held that vultures actually do not fly very high and that, in any case, odorous air being lighter than non-odorous, rises to levels where they soar. In support of his claim that vultures do not eat fresh meat he told of a large dead serpent hidden in a wood round which none of the birds assembled until, after three days, putrefaction had taken place; of scraps of fresh meat thrown out but never taken until putrid; of thousands of human corpses resulting from pestilence in Malaya not attacked by vultures because buried soon, whilst others washed up by the sea after putrifying, were torn to pieces; and of a sick negro, lying in the open, attacked by ants but not by vultures. It is *quite certain,* he maintained, "that the vulture is directed to its food by its olfactory nerves coming in contact with putrid effluvium rising through heavier air from corrupted substances".

It was not long, however, before Waterton himself became the subject of criticism and it was pointed out (by Hunter) that his views were "at variance with the observations of Wilson, Humboldt and Azara, three of the most philosophic and observant naturalists that ever lived". Vultures, Hunter continued, frequently soar to very great heights and they *do* feed on live animals even in preference where they can be obtained and are, in some parts, the terror of the poultry yard.

In order to rebut opposition, Audubon requested his friend, the Rev. John Bachman of Charleston, South Carolina, to conduct experiments on the faculty of smelling in vultures. Bachman, one of the best practical ornithologists of the U.S.A. at that time, not only confirmed Audubon's findings but persuaded members of the staff of the College of South Carolina to watch the tests. These witnesses affirmed that "the vultures of Carolina *(Cathartes aura* and *Cathartes jota)*, ... devour fresh as well as putrid food of any kind, and that they are guided to their food altogether through their sense of sight, and not that of smell". Bachman showed that it was sufficient to cover carcases and offal with a frame not touching the food or with thin canvas in contact with the food, to prevent the vultures from discovering the putrid material and that they attacked a coarse painting on canvas of a skinned and cut-open sheep but failed to discover offal fifteen feet away under an open frame. He also observed that a blinded vulture died of starvation while putrid flesh was within its reach.

In the year in which Bachman published his defence of Audubon, Charles Darwin, during his voyage round the world in H. M. S. Beagle recorded in his diary an account of an experiment on condors which he undertook in Chile. Twenty or thirty of these birds *(Gymnogyps gryphus)* were tethered in a long row. Darwin wrapped a piece of meat in white paper and walked backwards and forwards carrying the package in his hand at a distance of about three yards. The birds took no notice whatever. He then threw the package on the ground within one yard of an old male which looked at it for a moment but then took no further notice. It was pushed closer and closer until it eventually touched his beak. Then the paper was instantly torn off with fury and, at the same moment, every bird in the row began struggling and flapping its wings. Darwin remarked that "the evidence in favour of and against the acute smelling powers of carrion-vultures was singularly balanced" and in this view he was supported by Hutton who claimed that Indian vultures, though guided to their prey by sight alone, finally locate it by its odour. These vultures eat both fresh meat and carrion. When Hutton hung a canvas bag containing a freshly killed dog in a tree, they at first ignored it but collected in numbers when the carcase became very putrid.

Vultures, it should be remembered, are acknowledged to possess most wonderful eye-sight. Indeed, according to Beebe, "there is little doubt that this surpasses even that of hawks and eagles, and probably represents the highest development of the power of vision of any living creature". This claim, it is true, has also been made for wild turkeys which, however, seem to have but a dull apprehension of scent, a flock having been observed by an American hunter, Caton, to come within fifty or sixty feet of him without showing the least alarm.

European hunters or, at least, those who lived in the Pyrenees, were convinced that the vultures of that region must often be attracted by smell to the dead animals which serve as their prey since, in many cases, they could not possibly see the carcases beforehand. These hunters (and the shepherds of the region) sometimes know more than do the scientists, wrote De Bouillé in 1875. He apparently had not read the publications of the American and British observers and contented himself with relating the experiences of others. Those who defend the 'vision' theory, he maintained, can only *suppose* but not *prove* the absence of a sense of smell in vultures whereas the supporters of the "smell" theory can prove that sight cannot be the effective agent in some cases at least. And he added that since birds may be able to perceive an odour not perceptible to man, emitted by dead animals, it is immaterial whether the carcases are putrid or not.

Some years later, an interesting observation by Rhoads lent support to De Bouillé's view. Rhoads noticed that vultures collected in large numbers where ground covering the burial place of a cow and a horse had been ploughed up. The carcases had been buried several years previously and the observer could detect no odour. His deduction was that "the olfactories of a turkey vulture *(Cathartes aura)* can alone serve its purpose in the discovery of food".

In the United States, as in Europe, the controversy continued about this time. There, Hoxie noted that vultures failed to discover the stinking carcase of an ox while it was hidden by thick bushes but assembled in large numbers to consume it when it was exposed to their view. Two young turkey vultures which he reared, completely ignored strong-smelling offal hidden from their sight by newspaper on which they stood. Similarly, a young caged black vulture failed to perceive until he saw them, a putrid crab, the stinking carcases of herons, or the still warm entrails of fowls. On the other hand, forty or fifty vultures flew from a distance of half-a-mile to feed on the carcase of a black vulture as soon as it was thrown out of a tent where it had just been skinned.

Neither this impressive evidence against olfactory power (despite the presence of an olfactory system) in birds, however, nor earlier weighty evidence convinced Sayles. Can anybody show, he asked, any such non-functioning systems in any other species? Audubon's vultures, he asserted, were merely misled when they attacked the stuffed deer. They did not search for carrion *under* the canvas after consuming what was visible because they believed that no more remained — just as they would not eat the putrid soil on which carrion had lain. Darwin's test was inconclusive because he did not report the direction of the wind. Audubon, Bachman and Darwin were all wrong, he claimed, and he told (thirteen years after the event!) how vultures collected, from a distance of two miles, round a very putrid heap of hog offal.

But Sayles, in his turn, did not remain unanswered. He was very soon accused, by Barrows, of simply adding more to the already long list of fallacious 'proofs' and of using chance observations in an attempt to controvert the reports of careful observers. Barrows denied the validity of the argument from 'structure to function', quoting evidence against it. Did Sayles, he asked, rely on memory or on notes when he reported the thirteen-year-old observations? Why did he omit so many important details? Was he certain that the dogs which fed on the offal did not drop some of it in full view of the vultures? Why, also, did the vultures abandon their search after two hours if they could still scent the putrid material?

Still another believer in the olfactory powers of the vulture was Professor Carlo Fabani whose views were expressed in a communication to the *Rivista Italiana di Scienze Naturali* dated 1896. He was familiar with the opinions of Audubon, Darwin and others but he concluded without reporting any experiments or any personal observations that the sense of smell is keen in the vultures, moderately acute in ravens and aquatic birds, feeble in fierce carnivorous birds and absent or almost absent in insectivorous and graminiverous birds.

Birds of Fabani's last class were the subject of a very short report by Reeker to the Westphalian Provincial Society for Science and Art in the year 1899. He told how a colleague scattered food on the snow before his house and observed that birds (blackbirds, starlings, tomtits, chaffinches) which had been consuming the food and had been temporarily driven away by a cat, returned to comsume all the remaining food except a potato that had been bitten by the cat which, from the nature of the locality could not have been seen by at least some of the birds. The assumption was, of course, that the scent of the cat, adhering to this potato, repelled the birds.

In ornithological observations of the olfactory sense, birds of prey have naturally received most attention. Other birds, however, have not been neglected and Raspail, one of the earlier and one of the stoutest defenders of the view that they have a keen sense of smell, made his observations (in the closing years of the XIXth century) on pheasants, partridges, wood pigeons, titmice, rooks, magpies and blackbirds. He found that pheasants partridges and pigeons would readly come near a spot where he was concealed if the hiding place were down wind but not if it were up wind; that breeding pheasants appeared to be able to smell water at a distance of 180 metres; that hungry partridges walked considerable distances in snow to feed on grain which they had apparently discovered by smell; that titmice similarly discovered cheese in traps; that rooks, magpies and blackbirds dug larvae out of the ground without preliminary searching. And he affirmed that turtle doves abandon their eggs if they are touched by human hands, supposedly because of the odour imparted.

Whatever the value of Raspail's findings, it is noteworthy that he satisfied himself with observations. He conducted no careful experiments and applied no checks. Not a few of the publications on the olfactory sense in birds suffer from this defect.

About five years after Reeker and Raspail, Gill published a brief note on his observation of a caged condor. While dissecting a rabbit, he tells us, he worked at an open window about ten yards from the cage and quite out of the bird's sight. During the period while the brain of the rabbit was being removed, the condor remained quietly basking in the sun. But when the abdominal cavity was opened and the characteristic, though not powerful, odour of the warm viscera escaped the bird became intensely excited and remained so until, after half-an-hour, the window was closed. Commenting on this event, Gill states that he felt it impossible to doubt that the condor, having at once perceived the odour of the viscera, must have possessed a sense of smell of very considerable acuteness.

Although publications on olfaction in birds are not numerous, it is remarkable that, during the period 1905—1922, *Nature* (volumes 71—110) contained six or seven communications on this topic. The first, by Hill, begins by pointing out that whether or not birds can smell, they certainly possess "olfactory" organs. Hill was of the opinion, however, that these organs are not very effective and he was confirmed in this view on hearing from Guillemard that vultures had frequently been observed to be quite unaware of the existence, within a few yards of them, of the disembowelled carcase of a wildbeest or other game merely hidden in the hole of an "antbear". Nevertheless, he decided to test the smelling powers of a graminiverous bird, choosing for this purpose, the turkey. Two of these birds, lightly fed, were each given access to two heaps of food (grain or bread), one heap of each pair being strongly scented with material having a powerful odour. Asafoetida, anise, oil of lavender, zinc valerianate, camphor, acetylene, carbon bisulphide, hydrocyanic acid and chloroform were used. In no case was any evidence obtained that the birds perceived the odours. They showed neither preference nor repugnance. One of them even fed on grain in a sieve beneath which hydrocyanic acid was passing until he exhibited marked signs of poisoning. Although, as Hill pointed out, there is no reason for supposing that an animal will enjoy an odour which pleases us or dislike one which we find disagreeable, it is difficult to credit animals which appear unable to detect such injurious substances as chloroform and hydrocyanic acid with an effective sense of smell. By way of contrast, Hill remarked that his dog, which appeared to be almost indifferent to carbon bisulphide, showed great repugnance to these two substances.

Hill's report appeared at about the same time as that of Rouse whose results, though few and subject to correction, are of particular interest because he actually measured the effect of odorous substances (turpentine, ammonia, asafoetida, bergamot oil, lily of the valley) on the respiration of ten or fiften captive birds (pigeons). He showed that some of these substances occasionally quickened and deepened the respiration possibly because of an irritating action on mucous membranes.

At the end of his communication, Hill had expressed his desire for information about the habits of wingless birds, the mode of live of which more or less resembles that of a terrestrial animal and he quoted the opinion of Owen that the acuity of smell of the kiwi would be expected to exceed that of any other bird. The kiwi is unique amongst birds in respect of the relative size of the olfactory lobes in the brain and the great size of the olfactory capsule in the skull.

Eighteen months later came Benham's account of the olfactory sense in the kiwi. When earthworms covered with four inches of earth were placed in shallow buckets and offered to a kiwi, it became quite excited and probed to the bottom of the buckets for the worms. Since the bird is believed to be nearly blind, since the sniffing noise which it makes would presumably prevent it from hearing the worms, and since it was found that it took no interest in a bucket containing earth but no worms, it was concluded that the worms were detected by smell. Repetition of the experiment with dead instead of live worms gave the same result — one bird "would not even look" at a bucket containing earth only which it had searched on the preceding day but, when a couple of dead worms were hidden in the earth, went to work promptly "as if he knew worms were there". In still another experiment, a kiwi which had been trained to eat meat was similarly found to be able, apparently, to detect its presence by smell and would wake up, if asleep, and reach in the right direction if meat or worms were thrown near it. Benham felt that more crucial experiments were needed but provisionally concluded that a certain amount of evidence had been afforded for the existence in the kiwi of a keen sense of smell Such experiments, however, have apparently not been made but, a few years later, Strong applied a test to kiwis in the London zoo, using earthworms. He obtained no evidence that they had an acute sense of smell.

In the year 1908, Schwartz conducted some interesting experiments on caged birds (yellow-hammer, greenfinch, canary, chaffinch, linnet, goldfinch, siskin and others) which were offered, as their only food during the period of test, ten sorts of seed having physical properties very similar to those of seeds favoured by the birds and having odour and/or taste for man. In addition, three sorts of seed poisonous to man were offered. Aqueous extracts of some seeds rejected by the birds added to seeds which they favoured did not affect consumption of the latter but the residual extracted seeds were *not* consumed. No definitive evidence of the existence of a sense of smell was found.

A very similar conclusion, concerning vultures, was reached at this time by Beebe, the curator of birds to the New York Zoological Society. He maintained that these birds, although strictly carnivorous, are unable to kill their prey and hence must wait for others to kill it or for it to die. "It has been proved conclusively", he continued, "that they find their food by the sense of sight alone and indeed apparently lack the sense of smell". Nevertheless, he "carefully tested the power of scent in the South American and Californian Condors and the King Vulture and, if present at all, it is very slight indeed. In the Black Vulture, the sense is appreciable but even here it appears to function but little but it reaches a greater degree of development in the Turkey Buzzard". In a test on turkey buzzard and black vulture he caused them to fast for several days and then allowed them to have access to three identical boxes placed on the ground, one of the boxes only containing tainted meat. Although one or two of the birds apparently noticed the odour, none of them found the meat.

Much more elaborate were the experiments on hybrid ring doves of the ornithologist Strong who also had the advantage of having studied the unique collection of bird brain material in the Senckenberg Neurological Institute at Frankfurt a. M. After making some unsuccessful tests in which cedar oil, asafoetida, bergamot oil, clove oil and hydrogen sulphide on filter paper or rags or in bottles were held near the heads of ring doves, crows, common fowls, turkey vultures and a paroquet, he constructed a labyrinth consisting of four compartments opening into a central square enclosure from which any food placed in the compartments could not be seen. The labyrinth was provided with an air exhaust and with a device by which a current of air, free from or containing odorous material, could be passed into the compartments. Preliminary tests with white rats placed in the central enclosure showed that these animals appeared to have no difficulty in locating the source of odour which was passed into one only of the compartments — the one which contained food. Four hybrid ring doves were trained to search for food in the four compartments. Then, during a period of nine months they were tested twice daily to find whether they would detect food by responding to an odorous air current emanating from one only of the compartments, the one which contained the food. The odorous materials used were: animal musk, eau de Cologne, violet sachet and, especially, bergamot oil. Some evidence was obtained that the doves responded to bergamot oil but Strong did not feel able to say more than that "the behavior of some birds at least may be affected by olfactory stimulation" — an opinion which was soon echoed by the zoologist Hesse.

Birds, he stated, have no appreciable power to scent and he believed that, since all odorous substances, when volatilized, have high specific gravity and sink to the ground, they are of no use as stimulants to birds. He added that birds lack skin glands and have little or no smell. In the opinion that, because odorous substances do not reach the heights where vultures soar, these birds depend on vision rather than on olfaction he was supported by Reiser who further maintained that vultures assemble near dead animals *before* they begin to decompose and to emit the odour of carrion. On the basis of twenty-three years' observation in countries where vultures abound, Reiser stated that they act in concert, quartering the heavens, communicating to each other the news of their discoveries and also watching the movements of other animals that are likewise searching for food.

Not less wide was the observation of Chapman who, although disclaiming expert knowledge, obviously acquired very great experience with many varietes of birds (vulture, kite, eagle, woodcock, snipe, golden plover, seagull, rook, wood pigeon, grouse, raven, guinea-fowl, starling, blackbird, thrush, ousel, robin, wagtail, sandpiper, black-headed gull, jackdaw, blackcock, curlew, peewit) in various parts of the world, particularly Africa and Spain. His view was that the only birds which habitually exercise the olfactory faculty are ducks and geese. In other birds the sense is dormant or obsolete. It is quite impossible, he thought, for high-soaring birds to detect odours. These birds, moreover, allow the near approach of man even when the wind blows to them from him: they use their keen sight and acute hearing. Chapman did not affirm that some or all of the birds (other than ducks and geese) which he observed *cannot* smell but only that they do not use their olfactory power in food-finding. Of the ducks and geese he wrote, however, "there can be no sort of mistake or possibility of error". *They* have remarkable powers of scent and they ceaselessly rely on them, taking flight immediately when human scent reaches them. Conversely, they fail to detect the scent when the wind is against it. While admitting that there are many exceptions and puzzling cases he repeated that "with the single notable exception of the wildfowl tribe, the use of the olfactory organs is practically non-existent or, at least, obsolete from desuetude", and he regarded as significant the fact that ducks and geese normally remain at low levels where "scent travels horizontally".

It is perhaps not surprising that experiments and observations on the sense of smell in birds have seldom been made on certain species. Accordingly, an isolated report on the Australian laughing jackass *(Dacelo gigas)*, a bird of the kingfisher family, is all the more interesting. In 1912, Pocock observed that three of these birds, as well as some individuals of other families, refused to accept as food, larvae of the Eggar moth, apparently because they disliked the smell, which, however, could not be detected by the observer.

Two other less familiar species, the humming bird and the ostrich, were examined a little later by Sherman and Gillespie respectively. Sherman provided bottles of sirup for the humming birds. Thy consumed this material readily but they did not appear to select when vanilla and lemon were added to separate bottles. The females only were attracted to the bottles. Gillespie, who was then director of the Scottish Zoological Park in Edinburgh, later found that a female ostrich, relying perhaps on her sense of smell, exhibited distinct preferences for certain kinds of sweet biscuits and would reject others immediately.

About this time, several bird-watchers included comments on the sense of smell in the reports of their occasionally long, pains-taking and uncomfortable studies. Near Salisbury, Penrose, who watched a pair of nesting stone curlews from a short distance (less than three yards, eventually), concluded that they might have detected him by smell; Miss Best and Miss Haviland, whose subject was a brooding grey lag-goose, deduced that it seemed to be made suspicious and frightened when the wind was not too high and blew from the observers' photographic tent towards the nest; and Von Liszt decided that a male sparrow whose behaviour he observed had a very finely developed sense of smell. It had a great liking for a certain kind of food and became very excited when this food was brought near even when the appearence of the food was much changed and when it was brought from different directions.

In view of the contradictory and inconclusive nature of the evidence and reasoning, it also not surprising that birds have been credited with 'occult' senses. Alternatively, it has been suggested that the so-called olfactory organs of birds really have quite a different function. As McKechnie thought, their purpose might be to detect fine differences in the strength, temperature and humidity of the air currents encountered during flight. This view need not be discussed here but it may be recalled that terns in which the nasal cavities are occluded with wax retain their power to return to their nests over an unknown course and from a considerable distance.

According to Beck, the 'occult' (but not supernatural) senses attributed to birds include a 'homing' sense, a 'food-finding' sense and, possibly, a 'mate-finding' sense. He records a case of the carcase of a freshly killed dog in a deep hole where it was practically invisible. Despite the fact that their nearest haunt was eight miles away, vultures quickly located the carcase which, since the temperature was below freezing point, presumably emitted little odour. It is evident, however, that no physiological basis for these hypothetical senses is known and the remark has very rightly been made that the idea of a sense has little meaning if divorced from that of an appropriate mechanism.

The communications to *Nature*, mentioned above, mostly suggest that the sense of smell is not well developed in birds. In his paper, on the other hand,

Williams strongly supports the view that they have a good olfactory sense. Writing of the red-headed and black vultures of Trinidad and Panama, he expresses the opinion that "there can be no doubt that these birds have an efficient sense of smell". They had been observed to collect in large numbers on and round a field which had recently been manured with liquid slaughter-house refuse and could be smelled a quarter of a mile away, although nothing resulting was visible to the eye. They quickly found the body of an animal placed close to the trunk of a thick tree, quite invisible from above. And they also collected near a plant bearing flowers having a strong odour. Williams adds, however, that he believes the congregation of birds in numbers is due to their watching the movements of their neighbours.

This last is the explanation also given by Meek in his comments on Williams' views. In the course of these comments, Meek refers to the "telescopic eyes" of vultures, eyes which are used, he believes, not only "to watch what is taking place over a wide range below them, but to note what their neighbours are doing. If one disappears the rest in turn fly to the region to find out the cause." While not denying that ducks, geese and other birds which feed for long periods on land and marsh have effective olfactory powers, he follows Chapman, a wildfowler with long experience, in holding that with few exceptions birds, especially eagles and vultures, have no sense of smell. In the Sudan, Chapman told him, vultures immediately discover a carcase and pick it clean but fail to find the rich supply of meat from the carcase removed a short distance and covered with branches to serve as human food.

It has already been remarked that the upper air in which birds spend much of their time is probably not a favourable location for odours compared with the air near ground level. Because of this and because vision seems to be predominant sense in birds (although some of them also have a very acute sense of touch) an anonymous writer in *Nature* in 1922 underlined McKechnie's view that their often well-developed 'olfactory' organs may actually serve some other, non-olfactory purpose. As regards vision, it may be remembered that, according to the experiments of Schmid, a falcon was able to detect at 1660 metres a moving decoy made of crows' feathers whilst another, at 1077 metres, detected pigeons seeking grain in a mowed field. The decoy could not be seen by the human eye assisted by a glass of six magnifications. (An incident providing further evidence of the ease with which a bird's vision may deceive it, was recorded by Palmer in 1923. He saw a turkey vulture approach and examine a toy rabbit made of white plush evidently mistaking this object for an animal.)

In their work on *Die Vögel Mitteleuropas*, the Heinroths state that owls resemble falcons in rejecting tainted meat apparently because its odour, which is perceived *via* the mouth and throat, is displeasing. They also say that ravens are quite anosmic and find hidden pieces of meat by chance only but they do not discuss olfaction in wild ducks such as those hunted by fowlers in the Frisian Islands. There, it is a very old established custom to carry pots of smouldering peat to mask the human odour during the catch, for the fowlers believe that their scent in very readly detected by the wildfowl. They also believe that the ducks recognize the odour of tobacco smoke and hence smoking is forbidden during the hunt. Dr. Nolte, of Kiel, spent some time accompanying the fowlers in order to study the olfactory powers of the ducks but was not able completely to confirm these beliefs. He concluded, however, that the frequent rains in the islands carry down odorous material which the ducks then perceive.

Nolte's report initiated a lively discussion in the Dutch jornal *De Nederlandsche Jager*. Those who contributed (or some of them, at least) felt that his evidence was inadequate and far from convincing. They pointed out that huntsmen often subscribe to superstitious opinions and that tobacco smoke does not differ sufficiently from peat smoke to be a suitable alternative. One of them (Agricola) told of a hen partridge which almost touched him and his dog before it detected him. Another (Van Dorssen) quoted extensively from his hunting records and showed that even when conditions were as favourable as possible (odourous grease on boots, perspiration abundant, close approach) for scenting by the ducks, they detected him by sight and hearing only. Their view was that birds, even ducks, have little or no sense of smell. Some years later, another Dutch writer, Jongens, expressed the opinion that ducks may be exceptions to the rule that this is so.

An even more emphatic disbeliever in the olfactory powers of birds was Soudek who reported his findings at the Tenth International Zoological Congress (1927) and published them also an *Biologicke Listy* (1925). On one occasion, he observed about fifteen turkey vultures in Arizona. Two decaying, stinking carcases of hens were left, one uncovered, the second covered with grass in a shallow ditch only 25 steps apart. The vultures saw the covered one being carried, saw the author's tracks and saw the hiding place marked with plucked grass. But they found the first only. His experiments, chiefly with pigeons, and partridges but also with domestic ducks and several other species at first suggested that birds can learn to recognize the odours of anise and xylol when these substances are added to grain. Later tests showed, however, that it was the slightly moist appearance of the corn containing xylol which enabled the birds to distinguish it from the specimen containing anise. In another long series of experiments, he showed that pigeons could not perceive the odour of rather strong solution of ammonia. They frequently drank or attempted to drink it and were obliged to "jump away horryfied". Similar results were obtained with pyridine, alcohol, ether and other odorous liquids. Moreover, although the birds could readily recognize colour (methylene blue) and shape (differently shaped jars of drinking water) they gave no response to the odours of animals of their own or other (dog, fox) species. Since his colleague, the ornithologist Karásek, during years of watching, never observed any signs of olfactory orientation in waterfowl, Soudek concluded that "all birds are anosmatic creatures".

In 1928, experiments with turkey vultures were once again made in the U. S. A. by Leighton and Lewis. In an area frequented by vultures, Leighton exposed a small stuffed deer, having first shown by allowing a dog to range the district that no carrion was near. During two days, only a few vultures approached. Fresh meat was then placed in a hole and covered with grass. One vulture only took some of the meat after it had been partly exposed by a crow. When very putrid meat covered with sackcloth was later placed in the open only very inconclusive results were obtained and Leighton gave his support to 'vision' rather than to 'olfaction'.

The tests carried out by Lewis in cold and in warm weather were similar. His bait included the carcases of a hen, a terrapin, an opposum and two cats. Sometimes this prey was left uncovered, sometimes it was concealed under a bag or box. Observation of the behaviour of the vultures which assembled — on some occasions they discovered the uncovered bait — led him to conclude that "in cold weather, when little odor is thrown off, sight is the vultures' chief means of locating food but that they have no difficulty in locating it by scent alone in warm weather when the odor is strong". Lewis also observed that vultures certainly sometimes eat the carcases of carnivorous in addition to those of other animals.

If birds can recognize and be repelled by animal and other odours, an explanation would be provided for the popular belief, mentioned earlier, that eggs which have been handled by human beings may be abandoned by the parent birds. Although it is possible that some species of birds may act in this way, it seems certain that ringed plovers do not. In 1931—32, Marples studied the nests of more than forty of these plovers and conducted experiments on them with dislodged, artificially coloured and dummy eggs. Finding that neither appearance nor weight was of importance to the birds, in the replacement of dislodged eggs, he tested the effect of handling and of odorous substances. No disturbance at all was caused by handling or by the odour of paint, plasticine or oil of orange "One must therefore conclude", he wrote, "that the sense of smell in the ringed plover is either very deficient or altogether absent, or that the odour of the egg, whether native to it or not, has nothing to do with 'egg recognition'." In this connection, it should be remembered that some birds, apparently, eat food which is poisonous to man or which has odour and/or taste repulsive to man — a fact which was stressed by Groebbels in his work "Der Vogel" (1932) and was confirmed by Ishihara in the course of his study of the anatomy and physiology of the avian olfactory apparatus. Like Audubon, Bachman and others, Ishihara observed that vultures failed to find a covered carcase although it was covered while they watched and although they must have seen in the grass the tracks of the person who concealed it. And, like Soudek, he showed that pigeons and parrots exhibited no repugnance to pyridine or ammonia. One of his pigeons seemed in no way troubled when its nasal openings were plugged with vaseline. He suggested that possibly birds pay so little attention to odour because they breathe through the mouth and swallow very rapidly.

The pigeons examined about this time by Zahn, on the other hand, were considered by him to have a sense of smell but he also found that redbreasts and hedge-sparrows could not be brought to respond to the juice of earthworms nor to distinguish between benzaldehyde and nitrobenzene or between amyl acetate and methylheptenone. His ducks, blue titmice, thrushes, greenfinches, redbreasts and hedge-sparrows could be trained to respond to olfactory tests but his jays and starlings could not. For blue titmice, he claimed, the 'minimum perceptible' for skatole is similar to that for man. In his very numerous experiments, which were carried out in special cages equipped with electric circuits so that shocks could be given as warnings when wrong selections were made, he offered food with and without odorous material (oils of wintergreen, anise, clove, rose and eucalyptus and coumarin, skatole, benzaldehyde, nitrobenzene, amyl acetate and methyl heptenone). Sometimes the food was treated with quinine or with concentrated solution of common salt. Not very many birds were tested.

Two other investigators who maintained that pigeons have an appreciable sense of smell were Bajandurow and Larin of the State Medical Institute in Tomsk. They constructed a device in which the head of the bird was enclosed in a glass dome. Odorous and odour-free air could be passed into this dome as desired. Using eau de Cologne, valerian, methyl salicylate and peppermint, they found that an olfactory conditioned reflex could be established. Electric shocks were given to elicit a non-conditioned reflex.

A novel approach to the problem of the sense perceptions of birds was made in 1936 by Salverda ter Laag who noticed that siskins are highly selective in their choice of water for bathing and drinking. Accordingly she conducted experiments on three or four of these birds, offering them pure water and water containing odorous or sapid substances (citric acid, *p*-ethoxyphenyl thiocarbamide, hydrochloric acid, pyridine, quinine, skatole, sodium acetate, sodium chloride, sodium phosphate). On the basis of what may be considered rather inadequate evidence, she concluded that the olfactory thresholds of siskins for pyridine and skatole are approximately at the concentrations of 0.005% and 0.000,000,000,1% respectively.

Even less well founded were the opinions of Wagner who maintained, after testing three caged redbreasts, that they rapidly learn to associate one of two odorous substances with a particular taste. His odorous materials were skatole and plant oils and his sapid substances, quinine and sodium chloride.

Starlings and sparrows, on the other hand, were found by Jongens not to be repelled from their nests by the odours of aniseed, peppermint or phenol. Like Nolte, he knew of the custom of fowlers of using smouldering peat to counteract human scent. While reminding his readers that birds may differ from man in their standards of good and bad in odours, he also felt strongly inclined to the view that birds, ducks included, have a poor sense of smell only. For a few days, he watched a blind curlew which, apparently, was unable to detect the scent of dog or man. His view of the possible olfactory standards of birds was later supplemented by Taverner who pointed out that both birds and dogs may simply disregard, as of no importance, odours which attract the attention of man; and by Bartsch, who gave an account of a carnivorous parrot which exhibited marked selectivity in its choice of food, evidently because of olfactory or gustatory preference.

It was from the Physiological Laboratory of the University of Amsterdam that Walter's paper came during the war. Entitled "Some experiments on the sense of smell in birds studied by the method of conditioned reflexes", it is one of the most careful and exhaustive studies so far made and, except for the investigation of Bajandurow and Larin, the only one in which the conditioned reflex method had so far been used in avian olfactory tests. Moreover, it includes a very valuable review of the literature and most interesting general discussion of the whole matter.

The subjects of Walter's experiments were male and female pigeons, parakeets, siskins and ducks. Some of the ducks were blindfolded and some were blinded. In some of the parakeets, the olfactory nerves were cut. Walter conducted training and feeding experiments, he employed electroencephalography, he made investigations on conditioned gastric secretion, he studied the relationship between respiration and the action of odorous substances and he used visual and acoustic stimulation and pain. His odorous substances, offered at intensities ranging (to human beings) from clearly perceptible to strong, included amyl acetate, eau de Cologne, benzene, toluene, xylene, methyl salicylate, aniseed, clove oil, phenol, citral, chloroform, ether, lavender oil, essence of jessamine, pyridine and fishmeal. Conditioned secretion of gastric juice, he found, was easily established with visual and acoustic stimuli but never with olfactory. His conclusion, already quoted above, was that the birds which he tested did not possess the faculty of smelling. He felt that experiments on other species should be carried out and he emphasized that "if apparently positive results are obtained they should meet with the requirement of being unobtainable immediately after transsection of the olfactory nerves". Finally, he pointed out that the possession of fairly well developed 'olfactory' organs is not an *a priori* proof of the possession of an olfactory sense.

Despite the great weight of negative evidence presented by Walter, positive claims have continued to be made by subsequent investigators although not on behalf of the same birds. Tasmanian swamp harriers, for example, were believed by Sharland to detect food and human beings by means of the sense of smell; Gillespie, after testing a few caged vultures and ravens with putrid horse flesh and observing tits which rejected fat having an unfamiliar odour, felt that ravens and tits may have olfactory powers; Frings and Boyd concluded from feeding experiments on fifteen caged adult bobwhite quails that they can develop a preference for feeding vessels to which, perhaps, a 'species' odour has been transferred by the birds themselves; and Hamrum, likewise as a result of experiments with bobwhite quails, thought it possible that they can learn to associate food with a specific odorous substance. The substance, in his case, was coumarin which seemed to have a repellant effect.

Other odorous repellants such as naphthalene and coal tar were not found by Dambach and Leedy, in cage and field experiments, to reveal olfactory ability in birds. On the contrary, these workers discovered that, under certain circumstances, healthy pheasants of both sexes will eat maize treated with these repellants. Indeed, in one of their field experiments, *untreated* maize suffered less from their depredations than did treated.

The valuable evidence obtained by Calvin, Williams and Westmoreland in 1957 was, however. in line with that of Walter and they, too, employed the method of reflexes. Their paired conditioned stimuli to movement from the white compartment of a glass-topped wooden cage to the black compartment were electric light and the odour of a cheap penetrating perfume. The unconditioned stimulus was electric shock applied through the brass rods which formed the base of the cage. They observed no sign of highly developed olfactory sensitivity in the domestic pigeon and no indication of olfactory discriminating power. And although, only two years later, Michelson affirmed that he had demonstrated the existence in pigeons of discrimination which could "be attributed only to the presence ... of olfactory stimuli", the fact that his somewhat obscure report omits several important details renders it unconvincing. His rather elaborate apparatus included a grain feeder, three plastic discs transilluminated in different colours, a delivery tube through which a controlled stream of odourless or odorous air could be supplied, an exhaust system and a device for providing a masking noise. Two pigeons only were tested fifty or more times with *sec*-butyl acetate and *iso*-octane as odorants.

As Gurney showed in 1922, the literature on olfaction in birds is largely anecdotal. It is, moreover, full of the contradictions which caused Pumphrey (in 1948) to write: "Most of the positive evidence brought forward by one authority has been criticized destructively by another" and to add, "we know even less about their [*i. e.*, the birds'] sense of smell than we do about our own".

The ornithologists of the Severn Wildfowl Trust "proved conclusively" (also in 1948), from the behaviour of ducks which appeared to be able to detect the scent of concealed watchers, that these birds have a sense of smell considerably better than that of human beings but they have not yet carried out confirmatory tests.

It has been said that "the nightingale returns from Greece not merely to the same country but to the same field and the self-same bush. The swallow takes possession of the same nest". However that may be, against the view that birds can smell there is a great mass of testimony, especially the testimony that they have little or no power of detecting the presence of an enemy even if approaching down wind, until he is seen or heard. Nevertheless, some very puzzling problems remain to be solved. How do birds locate grubs hidden beneath ground or behind bark? How do sea birds detect, even in thick mist, offal thrown overboard from fishing boats? Above all, how do vultures, apparently, know that a death has occurred in a house, gathering on the roof as if hoping to gain access to a corpse which they cannot possibly have seen? We certainly do not know.

Considerably more than a hundred years after Audubon, Owre and Northington of the University of Miami, Florida, conducted several hundred tests of the olfactory power of two turkey vultures confined in large cages. Pans, some containing dead leaves only, others containing, concealed under dead leaves, putrescent horse meat or putrescent horse meat smeared on dead chicks, were placed in the cages. In a second but much less extensive series of tests, the birds were offered a choice between fresh and putrid horse and chicken meat. The results of the first series clearly suggested that the food was located by smell and those of the second indicated that the vultures preferred fresh to putrid meat and chicken flesh to horse flesh. When, after 18 months captivity, these two birds were set free in a region where no vultures could be seen, several other vultures joined the released pair within a few minutes. It should, perhaps, be added that Frank M. Chapman, curator of birds in the American Museum of Natural History, had earlier (1938) concluded from his lengthy observations on the island of Barro Colorado, a nature reserve in Gatun Lake (Canal Zone), that he had "proved conclusively that ... the turkey buzzard's nose served him even better than his eyes" and placed beyond question the importance of this bird's sense of smell in discovering food. He had watched and photographed the buzzards which collected at points where cadavers were hidden, these points having been completely ignored (for about a year in one case) by the birds so long as they contained no meat.

It is certainly true, as Betsy Bang subsequently showed, that turkey vultures have large and heavily innervated olfactory organs. And she considered it "probable that birds possessing large olfactory organs may respond instinctively to chemical stimuli, either alone or in complex interaction with auditory, visual, vibratory and other organs of special sense".

[It is interesting to reccord in this connection that Miller, making systematic observations of black footed albatrosses on a cruise in a research yacht belonging to the Scripps Institution of Oceanography, found some evidence that these birds responded positively to the odour of animal fat (especially bacon fat) and were apparently strongly repelled by that of turpentine-linseed flux in paints used for marking the birds.]

Amongst recent studies of olfaction in birds, that of Neuhaus whose excellent work with dogs has been reported above, is of particular importance. The birds tested were geese and pigeons, a tame robin redbreast, a titmouse and a thrush. As odorants, butyric acid, eugenol, amyl acetate, clove oil, skatole, ethyl mercaptan, formaldehyde and other substances were used. Many of the experiments were conducted in a three-compartment apparatus in which only one compartment was odorous. Although more than 3000 tests were made, Neuhaus was unable to teach the birds to respond to olfactory stimuli. Seemingly positive results were sometimes obtained but these could be attributed to optical perception. However, geese which were tested in a darkened, silent room showed, by changes in respiratory rhythm that, they perceived odour.

These results were reported at the first Wenner-Gren symposium on olfaction and taste (1962). At the second (1965) Bernice Wenzel, in a most valuable contribution, pointed out that one reason why conflicting results are so often obtained is that the wrong question is asked. Instead of "Can birds smell"?, we should ask: 1. can they perceive olfactory stimuli; 2. do they naturally regulate any aspects of their behaviour in terms of olfactory cues; and 3. can they *learn* to regulate such aspects in those terms? And we should remember that 1 could be answered affirmatively while the answers to 2 and 3 might be in the negative. Her own work and that of other recent workers indicated that affirmative answers are being obtained to all three questions.

These recent workers include Stager (1964), Tucker (1965), Fink (1965) and Henton and co-workers (1966). Stager conducted field experiments, free from all visual food stimuli, with turkey vultures and concluded that they have a well developed sense of smell. He used devices that introduced odours of decomposing animal tissue and of volatile chemicals into predetermined air currents. He also obtained evidence that olfaction does not play more than a minor role in food location by some other species of birds. Tucker, who is convinced that the olfactory systems of birds are functional, obtained electro-physiological evidence suggesting that the olfactory receptors of several species (*e. g.*, hen, goose, sparrow-hawk, black vulture) respond to the odour of amyl acetate and of other odorants. Fink adduced very convincing results showing that, although the olfactory organs of at least some birds are fully capable of functioning and that they do perceive odours, olfaction nevertheless of very little importance to them possibly because their olfactory nerves and brains are so constructed as to prevent, or greatly diminish, ability to form associations. And Henton and co-workers supported Fink's conclusions but used a technique involving a surgical operation on the very small number (three) of pigeons tested with amyl acetate.

Fishes that Tipple in the Deep

> Je me propose d'établier, dans le cours de ce Mémoire, que l'organe du goût n'existe pas et ne pouvoit pas même exister dans la bouche des poissons, par une suite nécessaire de leur manière de respirer; que les organes, regardés jusqu'ici comme propres à l'odorat dans ces animaux, sont destinés à percevoir une sensation analogue à celle des saveurs; enfin qu'il ne peut y avoir de véritable odeur pour un animal habituellement plongé dans un liquide. *Duméril*

> When the anterior nasal apertures of this fish *[Fundulus heroclitus]* are closed by being stitched up, the fish reacts to wads containing meat as it does to those without meat. On reopening the nasal apertures by loosening the stitches, the ability to distinguish between the two kinds of wads is again shown by the fish. As the power to discover distant hidden food is lost by both kinds of fishes *[Fundulus heteroclitus* and *Amiurus nebulosus]* when their olfactory organs are prevented from acting, even though their organs of taste are in normal condition, it is concluded that their olfactory organs are as truly organs of smell as are those of the higher vertebrates, and that their olfactory organs are, therefore, properly classed as distance receptors. *Parker*

> Man hat viel darüber gestritten, ob nur gasförmige oder auch flüssige Substanzen gerochen werden. Dieser Streit ist eigentlich müßig, denn da die Nervenendigungen mit einer capillaren Schleimschicht überzogen sind, muß sich jedes Gas zunächst in diesem Schleim lösen, ehe es mit den Nervenendigungen in Berührung kommt. *Zuntz*

Mammals living in water *(e. g.*, dolphins, porpoises, seals, walruses, whales) have poorly developed olfactory organs and are said, although the evidence appears to be scanty, to have an extremely poor sense of smell. As for fish, Duméril was evidently quite sure that they are completely lacking in olfactory power. "There can be no true sense of smell", he affirmed, "in an animal habitually immersed in a liquid." Parker, on the other hand, was obviously equally sure that the olfactory organs of fish are true distance receptors and are not less truly organs of smell than are those of the higher vertebrates. There exists also, of course, a third possibility — perhaps this was all that Parker claimed — namely, that some species of fish have a sense of smell whilst others have not. The dogfish, we are told, seeks its food exclusively by scent whilst the carp's search is wholly guided by sight. Thus, as in the case of birds, there has been a conflict of opinion. This, perhaps, has not been so sharp as was the ornithological conflict nor has any Audubon entered the lists.

Scientific study of olfaction in fish seems to have begun in the early years of the nineteenth century when French investigators turned their attention to the subject, Duméril in his *Mémoire sur l'odorat des poissons* (1807) and Geoffroy St. Hilaire in his *Mémoire sur la structure et les usages de l'appareil olfactif dans les poissons suivi de considérations sur l'olfaction des animaux qui odorent dans l'air* (1825). These are valuable and interesting publications. But the observations recorded in them are almost entirely comparative and anatomical and, if the authors actually carried out any olfactory experiments, with fish, they have omitted to mention them. Duméril, it is true, admitted that fish are attracted by various kinds of bait and by many materials which are strongly odorous to man. But he appears to have believed that they detect such substances by taste or, at any rate, by some sense not comparable to the human olfactory sense.

This was certainly not the opinion formed by Bateson when, more than eighty years later, he studied fish at Plymouth. Although he concluded that "the majority of fishes seek their food chiefly, if not entirely by sight", he likewise held that "a certain number hunt for and recognize it by the sense of smell alone". Fishes of this second class (*e. g.*, eels, rockling) show consciousness of food when it is unseen and they habitually seek food without the help of their eyes. Tests with blind fish and with rockling and congers in which the olfactory organ had been destroyed, confirmed this view. In short, said Bateson, "the perceptions by which these animals recognize the presence of food are clearly obtained by means of the olfactory organs and apparently exclusively by them". Besides recording interesting information about those fish (the majority) which rely mainly on sight, he also emphasized the difficulties connected with the study of aquatic animals and pointed out that his observations were entirely confined to fish living, not under natural conditions but kept in tanks.

The fish studied by Professor Baglioni of Rome, who conducted the next investigations to be considered, were likewise kept in tanks. The only species he used was the teleost *Balistes capriscus.* and all the individuals tested were blinded. The manner in which they detected the food he provided convinced him that their olfactory powers were perhaps as great as are those of land animals.

It was about this time (1907—1909) that American workers, in particularly Parker and his co-workers in association with the U. S. Bureau of Fisheries, began their studies on the sense reactions of fishes, studies which lasted for more than a decade. The fish examined by these researchers (dogfish, catfish, killifish, puffer) were kept in tanks or pools. Earth worms, extract of earth worms, living crabs, crab meat, beef, liver, frog muscle, propylamine and other amines were used as open or concealed olfactory stimulants. In addition to normal fish, blinded fish, fish with the olfactory tracts (forebrain) destroyed and fish with nasal cavities plugged were tested.

All these experiments clearly showed that some of the species at least find their food by smell.

For some time after 1918, interest in the olfactory powers of fish seems to have shifted from the U. S. A. to the European Continent and to Germany in particular although investigations continued in Canada and workers in Belgium, Britain and Finland shared in the experiments, the more important of which may now be considered.

In 1924, Strieck conducted training experiments which showed conclusively that his fish could detect and distinguish between skatole, coumarin and musk. He showed also that they have a sense of taste and can distinguish between sweet, sour, salt and bitter. In his view, the distinction between taste and smell is not merely a matter of the concentration of the substances concerned but depends rather on the chemical nature of these substances. Strieck studied blinded minnows only.

Pipping, on the other hand, examined several species including pike, flounder, and stickleback. The evidence which she obtained again indicated that some species depend chiefly on sight, others chiefly on smell in the search for food. Her most important finding, however, was that the olfactory sense was acute only in those species in which a constant current of water passed from front to rear through the nasal cavities.

Confirmation of this finding was soon provided by Wunder who examined, amongst others, carp, perch, and trout; and, a little later by Van Den Berghe who used Indian ink as a 'tracer' in order to follow the currents.

During this period, (1928—1930) Bull carried out some experiments with blennies using a synthetic musk as odorant and concluded that they respond to gustatory, visual, and auditory stimuli but not to olfactory.

In 1932, Wrede approached the problem of olfaction in fish from a different angle. Pointing out that minnows, removed from water, have (for man) a characteristic smell, she asked herself whether this smell has any significance for the minnows themselves. Although her training tests seemed to give a positive answer to this question, it also appeared that senses other than the olfactory were involved since the response was not prevented by removing the forebrain.

Still another, and no doubt less plausible approach, was that of Von Schiller of Budapest who likewise studied minnows. Having adopted Von Hornbostel's theory of 'bright' and 'dark' odours he studied minnows, employing musk ('bright') and indole ('dark') as odorous materials. His minnows were exposed to bright light and to darkness. He claimed that after light they preferred the 'bright' odour and after darkness the 'dark' odour.

Much more acceptable are the most important findings of Göz who also studied blinded minnows. His observations described in a very long paper in

1942 led him emphatically to conclude that fish both taste and smell and that they distinguish, by means of smell, between their own and other species of fish. Moreover, they detect not only 'species' odour but also 'individual' odour. When their olfactory powers were destroyed by operation, they could still detect 'species' odours but could not be trained to distinguish between them. Göz maintains, further, that human beings can distinguish the characteristic 'species' odours of fish.

Of all the Continental studies of olfaction in fish, the most interesting, and perhaps also the most important, are those of Neurath who published the account of her work in 1949. Like previous investigators, she used blinded minnows which she trained by 'positive' and 'negative' tests, to phenylethyl alcohol and eugenol, two substances which are *tasteless* to man. When their forebrains were removed, these minnows could not be so trained. Otherwise, however, their olfactory training was successful. It revealed the most astonishing fact that they could detect eugenol at a dilution of one part in 17 million and phenylethyl alcohol at a dilution of one part in 23 million. Thus, for these compounds, the olfactory acuity of minnows is 150 to 250 times as great as is that of human beings. As we shall later see, this finding acquires added significance in connection with the homing problem, that is to say with the problem of whether fish such as salmon which migrate from river to sea and later return to river, actually recognize their 'parent stream' in some way and return to it.

This question has exercised the minds of fishermen for many years and, on several occasions, attempts have been made to follow the course taken by tagged salmon in the sea. One of these attempts was made, in 1926, by Craigie whose observations were made in the waters near Vancouver Island. Although his findings were rather inconclusive, he felt able to state that "elimination of olfactory sensibility appears to interfere to some extent with migration", a conclusion rather similar to that of Williamson who, while convinced that salmon do indeed have a homing instinct, later expressed his opinion thus: "... a homing instinct discriminative enough to enable a salmon that comes, for example, under the influence of a river to recognize that that stream is not its native river is difficult to realize." Despite this conclusion, it was possible for Walker and Hasler, in 1949, to show that blinded minnows can be trained to discriminate between the odours of various aquatic plants and that, when 'fatigued' by the odour of one species, they can still detect the odours of other species. When the olfactory epithelium is destroyed, power of discrimination is lost. Here are their own words: "Results of our studies lend support to the view that aquatic plants may well play an important role in the life of a fish. They may serve as signposts to guide fish into feeding grounds, since many fishes commonly feed in turbid water, at dusk, at dawn and at night when visibility is poor. Moreover, the odours of aquatic plants may serve as attractants to immature fishes to prevent them from straying from cover. In addition, other natural odours may direct migratory fishes in locating their homing areas."

Before discussing later work by Hasler and his co-workers, we may refer to the studies of von Frisch whose long paper on *Die Bedeutung des Geruchsinnes im Leben der Fische* appeared in 1941 almost at the same time as did that of Göz. Working chiefly with intact and blinded minnows in tanks and in open water, he showed that the skin of living predatory fish contains an "alarm substance" (Schreckstoff) which causes minnows to react with alarm, imducing them to take flight. Alarm substances, which are present also in mammalian skin (e. g., of seal and sea-lion) are perceived at very high dilutions (one part in several million parts of water). He stated that minnows can distinguish, apparently by smell, between males and females of their own species and that they flee from the damaged skin of their own species. (This holds also for salmon.) The substance also appears to be released when the fish die. The minnows respond to the alarm substance only if their olfactory apparatus is intact.

Blinded minnows were again the fish chosen by Hasler and Wisby for their experiments in detecting phenol which is a common pollutant in municipal water supplies. They found by a reward and punishment method of training that these fish could learn to discriminate successfully between phenol and *p*-chlorophenol at concentrations as low as 0.0005 parts per million (ppm) or less and that they could also distinguish *o*-chlorophenol from the other two substances at low concentrations. When their olfactory sense was eliminated, the minnows failed to respond to the test substances. It is interesting to note that human beings cannot detect phenol by smell at concentrations below 0.01 ppm.

For a considerable number of years, Hasler and his co-workers and other Canadian investigators have conducted similar experiments, sometimes with added odorants, in tanks and in streams in order to be able to answer such questions as "Do streams have characteristic odours?" and "Can fish discriminate between these odours, retaining them in their memory from youth to maturity?". They concluded that, despite considerable chemical and physical changes, some of which are seasonal, streams appear to retain their own specific odours for years and that minnows can retain these odours in their memories for periods of several weeks. Analysis of the waters showed that the material concerned was organic, volatile and possibly nitrogenous. Later, it was shown that when the human hand or water used to rinse it was placed in water flowing into an area where salmon were migrating up stream, the migration was suddenly and strikingly reduced. Accordingly, it seems that the human skin also contains an alarm substance. It has also been suggested that the preference of salmon for night as opposed to day movement may be connected with their aversion to human scent. Still later, it was found that many organic substances including shark repellent and pulp mill effluent do not check the migration. Chemical analysis of water used to rinse mammalian skin has since shown that the alarm substance, which acts at very high dilutions, may be the amino-acid 1-serine. Other amino-acids have no repellent effect.

Despite the very heavy mortality that occurs in the sea, it is known that remarkably high proportions (*e. g.*, 11,000 out of 470,000) of marked salmon return to their parent-streams (none to nearby streams) from distances of hundreds or even thousands of miles after up to five years (similar results have been obtained with sea trout) and it is with these fish that much of the most interesting work has been done. This work has shown that salmon return, not only to their parent-river system but even to their specific tributory and that the ability to do this is seriously impaired if their olfactory passages are blocked. Thus, salmon removed to a place one mile below the junction of their parent-streams showed excellent ability to repeat their original choice at the junction whilst other salmon in which the olfactory pits were plugged, exhibited no such power. Similar results were obtained with salmon making a return journey of forty miles in a stream having fifteen to twenty tributories. Even more surprising, perhaps, is the fact that salmon hatched in a particular stream and *flown* out to a river whence they migrated to the sea, returned to the stream in which they were reared and hence *used a route they had never before travelled.* Although it was scarcely needed, further confirmation of the significance of home-stream water for salmon was obtained by injecting it into a large aquarium. The injection caused prolonged excited movement of the salmon present.

It has long been known that the eel has a very remarkable life history. Eels apparently locate distant food by smell and migrating elvers discriminate between ebb and flood tide, having a preference for fresh inland waters based on olfactory stimuli. When these waters are treated with charcoal or diluted with seawater, they lose their attractiveness. It seems that elvers differ from salmon in that their olfactory power is probably inborn whereas, in salmon, the power is apparently acquired. However that may be, the remarkable findings of Teichmann, who conducted thousands of tests with intact and blinded eels, appear to show that, at least for some substances, their olfactory acuity is as good as or superior to that of all other animals so far tested, man and dog included, since they can detect as little as two or three molecules of some compounds. He showed that some of them can detect phenylethyl alcohol at concentrations of $1:(2.86 \times 10^{-18})$ or even less. Expressed in another way, this means that they can detect 1 c.c. of the compound in a volume of water 58 times as great as that of Lake Constance which is 30 miles long, 10 miles wide and is 827 feet deep at its deepest part. They can learn to distinguish between various odours (*e. g.*, those of citral, eugenol, ionone, menthol and terpineol which are detected at similar concentrations) and to associate them with punishment and reward. For natural odorous substances, crushed worms for example, their acuity is equally pronounced. Like human beings, they exhibit large individual differences in acuity and this is subject also to appreciable seasonal variations.

Several other species of fish *(e. g.*, catfish, goldfish, sharks and sunfish) have been tested (electro-physiological tests have been applied in some cases) with odorous biological *(e. g.*, beef, clam juice, cod muscle, sheep kidney) and chemical *(e. g.*, acetic acid, butyl alcohol, coumarin, nitrobenzene, vanillin) substances and with all or almost all species, it has been found that olfactory acuity is pronounced or very pronounced. Many species, it seems, locate their food, even at long distances, and avoid their enemies by means of their sense of smell. The importance of the concentration of the odorant is shown by the fact that, with some substances, a high concentration may repel while a low may attract. With goldfish, it is possible even to induce stereotyped feeding activity indistinguishable from normal feeding activity by electrical stimulation of the olfactory areas of the central nervous system but not by stimulation of other areas.

An entirely different, but not less important fundamental study of olfaction in fish was described, in 1938, by Adrian and Ludwig who made records of the electric response of the olfactory stalk of the catfish and other species such as carp, goldfish and tench. Definite olfactory discharges were obtained when the stimulants were aqueous extracts of fresh or decayed animal tissues *(e. g.*, earthworms, liver, blood) but there was no discharge when rose water, oil of cloves, milk, starch solution and some other materials were used.

From all the evidence given above, we may safely conclude, then, that at least some species of fish have a sense of smell; that again in some species, this sense is probably extremely acute; and, finally, that migratory fish even after long absence at great distances from their home areas, may be able to locate their parent-streams by means of their odours. The olfactory powers of fish, it seems, are indeed not less amazing than are those of some land animals. To the scientist and, no doubt, to the fisherman also, the knowledge so far acquired is clearly of the greatest interest. Presumably, too, further revelations will continue to be made in this most fascinating field and it is perhaps peculiarly noteworthy that, in the cave-dwelling characins of Mexico, the sense of sight, but not that of smell, has been lost.

Very recently (1965), Hara and his co-workers have shown by electro-encephalographic tests, that adult spawning salmon respond vigorously to water from their "home" pond but are indifferent to water from other sources and a little earlier (1963) Kleerekoper and Mogensen found that the parasitic sea lamprey *(Petromyzon marinus)* responds, not only to the scent of trout but also to the odour of either one of two amines obtained by chemical separation from that scent, one of those two having a much more powerful effect than the other.

Savage and Blind

> The general conclusion which may be drawn from the available evidence is that pure sense-acuity is much the same in all races in the absence of definite pathological conditions such as errors of refraction and ruptured ear membranes, and that the frequent superiority of the savage over civilised man in his recognition of what is going on around him in nature is due to his trained powers of observation, powers usually limited in scope but very highly developed in special directions. *W. H. Rivers*

There is a widespread belief that primitive peoples and blind persons (or, at least, persons born blind or blinded while young) naturally have a particularly acute sense of smell. Some travellers have even claimed that the olfactory power of the savage is equal to that of the dog. Thus, the power to follow the trail like a dog has been attributed to the peoples of the Torres Straits (Myers was unable to confirm this), to Australian Kanakas, to the Indians of Peru and other parts of South America, and to the kaffirs, whilst the maroon negroes of the Antilles were said to be able to distinguish their own tracks from those of a Frenchman by odour. Again, it seems certain that some blind people have a remarkably keen sense of smell. It appears, however, that those who attribute special sensory powers to the savage and the blind have usually forgotten to consider whether the hyper-acuity, if indeed it exists, is inborn or acquired.

As to the savage, some naturalists have noted that his olfactory power may not really be superior to that of the civilized man [1]. They have observed that it has been more with his eyes than with his nose that he has successfully followed the trail, and, recalling the extraordinary olfactory ability of the civilized specialist [2] *(e. g.*, the tea-taster) have pointed out that "what is rare amongst civilized men is common amongst savages. That is all".

As to the blind, similar considerations should apply but it must be added that when one sense has been lost the others have been credited with vicarious power.

---

[1] Colenso has reported that the Maoris have a very keenly developed sense of smell.

[2] The civilized specialists include wine-tasters. In 1907, Vaschide briefly reported some tests which he had carried out with two wine-tasters. The first of these, with his eyes blindfolded, made not a single error in determining the age, value, and taste of seventeen sorts of wine. When his nose was stopped and when he had a cold, he made mistakes and when liquors such as champagne and brandy were offered, he made gross errors. The second taster was not so accurate as the first in determining the age of wines. He correctly distinguished 33 sorts of Bordeaux wine, however, and made only one mistake in testing wines which had been diluted with various quantities of water. Like the first taster, he was tested blindfold.

The experiments to be described may not be conclusive. Nevertheless, they strongly suggest that neither the uncivilized nor the blind naturally have any special olfactory ability.

Savage

Apparently the first primitive people to be subjected to experiment were the Dinkas of the Sudan who were studied by Professor Cesare Lombroso and his co-worker Mario Carrara about seventy years ago. One odorous substance only, oil of cloves, was used. It was offered in dilutions ranging from 1 : 200 to 1 : 50,000. Three of the Dinkas tested failed to recognize the substance at any of the concentrations employed and nearly all of them could not recognize any concentration lower than 1 : 2000. Lombroso and Carrara give no details of their method nor do they record the number of subjects tested. It seems, also, that they made no comparative tests on civilized subjects. They pointed out, however, that language difficulties may have interfered with the accuracy of their results.

The next group to be examined were the Murray Islanders whose olfactory acuity, ability to discriminate odours and odour-strengths and power to remember odours were tested by Myers when he took part in the Cambridge Anthropological Expedition to Torres Straits.

He first used valerianic acid as test substance but, because of the extreme lack of permanence of the odour of its solutions, he abandoned it in favour of freshly prepared aqueous solutions of camphor, the smell of camphor being quite familiar to the Islanders. He also abandoned his original intention to use Zwaardemaker's olfactometer. It was found impossible to use this instrument under the conditions involved and, in any case, the Islanders disliked it. In its place, glass cylinders containing water and progressively diluted solutions were employed, the order in which these cylinders were presented being changed without the knowledge of the persons tested. On all occasions, members of the expedition were tested as controls and, at a later date, adults and children in Scotland were likewise examined. In addition to camphor, scents from Bond Street, London and other odorous materials were used, the subjects being asked to give their opinion *(i. e.,* to describe in terms of some odour with which they were familiar) as to the nature of the odours and to indicate like or dislike. All the Islanders took a very great interest in the experiments, "giving their answers with the utmost care and deliberation". (Some Australian natives were also tested at this time by Seligmann acting on Myers' behalf. He likewise tested some Western Islanders. The Australians showed much less intelligence than the Western Islanders and took far less interest in the experiment.) As would be expected, Myers encountered many difficulties. For example, he required odourless water but found that "in Murray Island everything seemed to have a smell". Like Lombroso and Carrara he was faced with problems of language. And, in particular, he was obliged to use somewhat unsatisfactory methods and to confine his examinations to an insufficient number of individuals. In spite of all this, he was able to conclude first, that the average olfactory acuity was only slightly higher in Murray Island and the Torres Straits generally than in Aberdeenshire and, second, "that the people of the Torres Straits have much the same liking and disliking for various odours as obtains among ourselves". Thus, the Murray Islanders, who gave their replies with surprising readiness and assurance, disliked asafoetida, valerianic acid and, especially, civet. Musk, thyme and, above all, violet were their favourite odours.

Soon after the studies of Myers were concluded, the Todas of the Nilgiri Hills in Southern India were examined by Rivers who employed methods as similar as possible to those used by the Torres Straits Expedition. Unlike the inhabitants of the Straits, the Todas knew no English and Rivers had to rely on the help of an interpreter. He believed, however, that this difficulty was partly counterbalanced by the greater intelligence of the Todas whose care and attention to the observations were admirable. The odorous substance used was camphor. About fifty Toda men, a few Toda women and, as controls, a number of English observers were the subjects of the tests. Rivers felt for various reasons, that it was useless to attempt any exact comparison between Todas and English subjects. But he remained fairly confident that the Todas showed no superiority in acuity of smell but were rather somewhat inferior.

The conclusions reached by the next worker whose investigations will be reported were different. Grijns, in 1906, compared some Javanese with some Europeans and reported that the former were about twice as sensitive as the latter. He used Zwaardemaker's olfactometer with aqueous solutions of acetic acid, phenol and ammonia. These can scarcely be regarded as suitable substances. Nevertheless, two of them, acetic acid and ammonia, were the only ones used fifteen years later by Koster in his comparison of Dutch persons with natives of the Netherlands East Indies. Koster, who had no olfactometer at his disposal, used strips of filter paper impregnated with the odorous materials. His results, like those of Grijns, showed that the natives had a keener sense of smell than had the Europeans.

Although the available evidence concerning the olfactory acuity of primitive peoples is thus contradictory, readers of the works summarized above will probably feel that the Dutch investigators fail to convince especially because of their unfortunate choice of compounds used as test substances. They fail also because it seems evident that the other senses of savages are likewise in no way superior to those of civilized man.

Blind

The olfactory sense of the blind seems to have been investigated even less frequently than has that of savages. Only four series of experiments appear to have been reported so far, the first by Griesbach in 1889, the second by Mahner in 1908, the third by Cherubino and Salis fifty years later and the fourth by Boccuzzi in 1962. (Henning stated that he did not find blind children superior to seeing children but he gave no account of his experiments.)

Griesbach's subjects were chiefly boys of a blind school but he also examined four blind girls one of whom was also deaf and one also deaf-mute. Seeing pupils from other schools were tested as controls, these pupils being of comparable ages and at the same educational stages. All his results pointed in the same direction. In olfactory power, the blind children were inferior to the seeing.

The opposite conclusion was reached by Mahner. With Sternberg's olfactometer, he tested four blind, four deaf-mute and four normal children, the deaf-mute and normal children being blindfolded during the tests. His odorous substances were lily of the valley perfume and potassium sulphide. Each child was tested twenty times with each of these substances at two concentrations, the number of correct, false and doubtful judgments being recorded. On the average, the blind children, with lily of the valley, made ten correct judgments, the deaf-mute 6.75 and the normal 6.0, the corresponding figures for the sulphide being 13.5, 7.75 and 5.75. Thus, not only the blind but also the deaf-mute children were found to be considerably superior in discrimination to the normal children, the deaf-mutes being inferior, however, to the blind. Similar results were obtained when the sense of touch and that of taste were studied.

In the much more recent studies, the results were distinctly more in agreement with those of Griesbach than with those of Mahner. Moreover, the investigators (Cherubino and Salis) tested a considerably larger number of subjects, men, women and children. Of their 172 subjects, thirty-one (twenty males and eleven females) were blind (two groups, ages 5—15 and 16—35 respectively) fifty-five were deaf-mute and the remainder were normal subjects of both sexes in the same age groups as were the blind and the deaf-mute. For the tests, a modified form of the olfactometer of Fortunato and Niccolini and the following substances were used: lavender, roasted coffee, citral, rhodinol, vanillin, pyridine, rose water and peppermint. No compensatory hyperfunction was found. On the contrary, there was a high proportion of hypofunction.

The conclusion reached by Boccuzzi, who used coffee, citral and vanillin as odorants and examined thirty-one persons born blind, sixty-five persons who had been blind for at least seven years and a hundred normal people was that, amongst persons having high acuity, the blind are superior to the seeing in the perception and identification of coffee and citral. Amongst persons with lower acuity, however, the blind do not differ from the seeing.

It should be noted that Mahner recorded ability to distinguish differences in concentration (intensity) whilst Griesbach determined thresholds. Both investigators examined rather small numbers of children. It is clear that much further work must be done and that modern knowledge must be applied before we know whether those who are deficient in one or more of the senses are naturally compensated by hyperacuity of the remaining senses. The evidence at present available suggests that they are not, although, as reports about Laura Bridgman, Helen Keller and others show, they, like those who are not deficient, may acquire astonishing sensory powers. (James Mitchell, the boy born blind and deaf, described by Professor Dugald Stewart in 1815, had an extremely acute sense of smell but, in a room, he could not discover a person twelve feet away by smell alone or, despite affirmations to the contrary, could he follow the trail of a person for two miles by smell.)

Law and Order

Law

> The peculiarly unsatisfactory character of the determinations of olfactometry and odorimetry is due chiefly to the fact that olfactory qualities, unlike visual and auditory, are not demarcated. It is true that it is more difficult to keep uniform the duration and extension of smell stimuli than it is to regulate these attributes for other stimuli, with the possible exceptions of temperature and taste. It is also true that the great gulf of psychophysics, our ignorance of the physiological processes which everywhere link the strictly physical to the psychological, is wider in the case of temperature, taste and smell than in the cases of vision, audition, pressure and strain. Yet, at best, the measurements of physics must always be in terms of sensation and the measurements of sensation must always be in terms of physics.
>
> *E. A. McC. Gamble*

> Since all knowledge comes to us primarily through the evidence of the senses, it is unavoidable that some of the greatest controversies which have occurred in the development of the sciences should have been centred around the *modus operandi* regulating the translation of a physical stimulus into the energy of thought.
>
> *W. Peddie*

Although the Weber-Fechner law, namely the rule that the increase of stimulus required to produce a just appreciable increase (the "just noticeable difference") in sensation must always bear the same ratio to the whole stimulus, has been described as the principal generalization of psychophysics, the attempt to discover whether or not it is applicable to olfaction has not often been made. As far as other senses are concerned, it is true, the law seems to hold, if at all, within very limited ranges only. Indeed, Rawdon-Smith goes so far as to call it "this essentially incorrect law". Nevertheless, it seems worth while briefly to review the work of at least some of those who have studied the question.

In his book, *Die Physiologie des Geruchs,* Zwaardemaker recorded that he encountered difficulties in testing the applicability of the law to smell. He noted, in particular, the disturbing effects of fatigue and of high concentrations of odorant. Finding that he obtained good results when using vulcanized rubber as odorous material, he concluded, perhaps on rather inadequate grounds, that as far as smell is concerned, deviations from the law are not great.

Zwaardemaker's first successor in this field was Eleanor A. McCulloch Gamble. No other investigator appears to have studied the question more carefully or more thoroughly than she. In fact, her's is one of the small number of full-scale investigations of the problem so far made. Her researches, as she claimed with not a little justification, "offer some evidence that Weber's law applies to smell" and that the value of the increase of stimulus required to produce the appreciable increase in sensation "lies between one-third and one-fourth". The method which she employed was that of just noticeable differences. She corresponded with Zwaardemaker, followed his suggestions, used his olfactometric method, and chose her odorants from his odour classes. These classes, however, she regarded as unsatisfactory and used them only on the principle that "even a most unsatisfactory system is better than none". Her chief object was to present the results of a mass of difference-determinations together with a statement of the controls to which these required to be subjected. But her paper[1] contains, in addition, a valuable discussion of many of the fundamental questions involved. In the first place, she emphasizes the difference between olfactometry, which is concerned with the keenness of the sense of smell and odorimetry which is concerned with the measurement of intensity of smell stimuli considered as objectively as possible. The indeterminateness of olfactory qualities, she says, makes all quantitative work on smell more or less unsystematic. We are obliged to regard as a simple and separate quality the odour of every substance which, from a physical point of view, is unmixed, although it seems unlikely that there are as many elementary odours as there are unmixed odorous substances. Discussion of the evidence in favour of this view leads to difficult but interesting questions, particularly about olfactory compensation, fatigue, the classification of odours and their relationship to tastes, olfactometers and olfactometry, the mechanism of

[1] American Journal of Psychology, 1898—99, vol. 10.

olfaction, the adsorption of odorous substances, anosmia and hyperosmia, and many other topics. No review can do justice to her paper which should be read in its entirety.

Gamble was soon followed by Toulouse and Vaschide who tested ten children, ten women and ten men using "decimal" solutions of camphor in water. Although they claim to have shown the law, applied to olfaction, to be "true in principle", the evidence they present is very slender and impossible to weigh because so lacking in detail. Their just noticeable differences appear to be very much higher than those determined by other workers.

Somewhat similar criticisms apply to the report made by De Sanctis to the Fifth Congress of Psychology in the first year of the present century. He used Zwaardemaker's simple olfactometer with cardboard cylinders impregnated with camphor from decimal alcoholic solutions and he also claimed to have demonstrated that the law is at least approximately true for olfaction.

In 1909, however, one of Zwaardemaker's pupils, Hermanides, determined the just noticeable differences for the nine "standard" odours of his teacher's classification. Although his results were not identical with those of Gamble, the degree of concordance was surprisingly high, his average value, 38%, being very near the higher of Gamble's two limits.

In their study of the intensities of odours and of the irritating effects of warning agents for inflammable and poisonous gases,, Katz and Talbert examined 55 substances and stated that they found Weber's law to hold throughout. However, they offered little or no evidence in support of this statement. Much the same criticism may be made of the work of Masuda whose results, published two years before those of the United States workers, were obtained with amyl alcohol.

Two other discussions of the law which were published before the second world war may now be mentioned. The first was published in 1933 in the *Journal of the New England Water Works Association.* In this discussion, the chemist of the City Water, Light and Power Department of Springfield, Illinois, reported that a table (presumably employed in connection with measurements of odour of drinking water supplies) used in the Springfield laboratories in which odour values increased by increments of 25%, was found, in a series of 86 determinations by various observers, to confirm the law. Apparently, however, the value was nearer to 50% than to 25%.

The authors of the second discussion were Zigler and Holway who used two simple "osmoscopes" somewhat similar in principle to Zwaardemaker's olfactometer and themselves acted as subjects. Their odorant was rubber and their conclusion was that "sensitivity and olfactometric strength are probably related by a function that is hyperbolic in form".

Since the war, five or six additional tests of the validity of the law have been made. Those workers who claimed that their results support it were

Wenzel, who found a "Weber fraction" of 0.15 by testing four subjects with phenylethyl alcohol; White whose odorant was coal gas; Blondal who chose the odours of citral, coffee and naphthalene, injecting them in regular short bursts into the noses of two subjects; Jones whose tests were conducted with aliphatic and aromatic hydrocarbons, aliphatic alcohols, ethyl acetate and pyridine; Reese and Stevens whose subjects, twelve men and eight women, sniffed air containing the odours of coffee and heptane; and Stone, the number of whose subjects was forty-eight. Stone used 2-heptenone in an olfactometer of his own construction and stated that his results conformed well to the law.

On the other hand, Le Magnen maintained that the law is not valid for olfaction and Mrak and his co-workers, who, like Wenzel, used phenylethyl alcohol which was smelled by fourteen subjects, found that they exhibited very great variations in ability to detect differences between the intensity of an unknown higher or lower concentration and that of a standard.

It should be said here that various modifications of the law have been suggested. Bachem, of the College of Medicine of the University of Illinois studied reports on the application of the law to the various senses and came to the conclusion that, although the original form of the law was unsound, some at least of the criticisms of it were unjustified. Account had to be taken, he felt, of central and peripheral effects and of the many neurophysiological elements involved. Bachem plotted an "integrated olfactory sensory curve" and decided that "the modified Fechnerian law represents a particular, complex dose-action law characterized by absolute threshold and quasi-logarithmic response and resulting from the interplay of many neurophysiological elements, *i. e.*, excitatory processes, distribution of receptor-sensitivity, responses of single nerve-fiber, synaptic summation, reflex adaptation, masking, central inhibition and suppression". However, it can scarcely be said that he was successful in applying the modified law to the sense of smell because he did not note any logarithmic and saturation responses and because he was forced to attribute the "exceptional behaviour of olfaction to peculiarities of its mechanism".

It is possible then, that the Weber-Fechner law (or some modification of it such as the proposed "power law" of Stevens) may eventually be found to be valid for olfaction. However that may be, it is also possible that the modifications of the law which have been suggested are not necessarily superior to the original form, and indeed recent exact experiments by Eisler, Stone and others have shown that the law may hold approximately for smell, the just noticeable increment being between one-third and one-fifth.

Stone and his colleagues have continued their studies of the law during 1964 and, having tested nine subjects with acetic and propionic acid, have concluded that the just noticeable increment (a little more than one-quarter) is indeed a constant fraction of the stimulus strength.

Order

> Mais il est permis de se demander si beaucoup de savants, mus par le besion de ce qu'il croyaient être la simplicité, n'ont pas établi à tort dans l'ordre des phénomènes le système des compartiments étanches, et s'il ne convient pas de se borner à constater la sublime et réelle unité du plan de la nature.
>
> *P. Kennel*

> N'est-ce pas une singulière manie de penser que les choses deviennent plus claires lorsque nettement énumerées et classées en categories distinctes? C'est devenue la coutoume chère à notre esprit, à laquelle il doit sacrifier sous peine de déroger ou de s'apparaître comme manquant à la clarté.
>
> *C. Regismanset*

In one of his most interesting articles, published more than forty years ago, E. J. Parry wrote as follows: "There have been numerous attempts at the classification of odours all of which have been empirical and useless and most of which have been childish and absurd[2]." And he then went on to suggest some considerations material to any possible scientific system of classification. The systems of Bain, Heyninx, Linnaeus and Wåhlin, Piesse, Rimmel, Zwaardemaker and others were almost certainly familiar to Parry. Since he wrote, not a few other systems, notably that of Henning, have been published. But it may be assumed that his contention cannot yet be seriously challenged.

The classifiers, of course, cannot have shared Parry's opinion. Not infrequently, they must have perceived at least some imperfections in their schemes. But, with one exception, they have not stressed this. The exception, as we shall see, was Fourcroy.

It is clear that, for many reasons, there are serious difficulties in classifying odours. Their number may be very large although Von Skramlik, who tested five observers with over two hundred odorants, maintained that there are only, very roughly, about fifty pure odours. (Pure odours are those which do not affect any sense except that of smell and which cannot be localized as between the two halves of the olfactory region.) Poucher, however, who made a classification based on rates of volatilisation of odorants, affirmed that a perfumer must have intimate knowledge of, perhaps, a thousand aromatic substances and hence, presumably, of something like a thousand odours. And Teuber, who examined reports on about 35,000 organic compounds, found that about three thousand were said to be odorous, about four hundred terms being used to describe their qualities.

---

[2] In this connection, it is most interesting to record that, at the December 1949 meeting of the Society of Cosmetic Chemists in New York City, Sagarin read a paper entitled *On the inherent invalidity of all current systems of odor classification.* In this paper, he discussed many systems, particularly that of Crocker and Henderson, showing that they all lacked reproducibility.

These, too, cause difficulties. For although there is general agreement about them — most people probably like floral and fruity odours and dislike odours of putrefaction — such agreement is not by any means universal. Many people enjoy the odours of decay in meat and cheese and Williams has recorded that he knew one person who liked the odour of skunk[3] and another who liked it when it was not too strong. Moreover, as Kenneth found, about forty years ago, when he offered twenty-six odorants *(e. g.*, amyl acetate, asafoetida, musk, vanillin) to twenty-nine male and thirty-four female subjects aged 18 to 24, likes and dislikes are appreciably affected by associations.

The remainder of this chapter is devoted to accounts of some classifications, old and new, presented with little or no comment.

About the time of the French revolution, two schemes were published in France. The earlier, that of Lorry, appears to have been devised so that a more exact knowledge of the action which medicaments exercise on our bodies by means of their volatile and odorous parts might be attained. His five classes were, the camphoraceous, the narcotic, the ethereal, the acid-volatile, and the alkaline.

Some fifteen years later, as we have seen, Citizen Fourcroy wrote his *Memoire sur l'esprit recteur de Boerrhaave, l'arome des chimistes français ou le principe de l'odeur des végétaux.* The system described in this paper also contains five classes, namely, the extractive or mucous odours, the fugitive oily odours, the volatile oily odours, the aromatic acid odours, and the hydrosulphurous odours. Fourcroy, as we have already remarked, was well aware of the shortcomings of his classification. "Mais, à coup sûr", he says, "cette division, cette classification est arbitraire, incertaine et vraiment caduque, puisque les impressions protéiformes de nos sens, et surtout de celui de l'odorat, n'ont rien de fixe, de permanent, d'égal, ni pour tous les hommes ni pour tous les tems, dans le meme individu".

In the interval between 1798, when Fourcroy's paper was published and 1907 whem Matteotti constructed his "natural" scale of odours *(Scala naturale degli odori)* several systems of classification were proposed — some, like that of Zwaardemaker, being particularly well known. In making his classification, Zwaardemaker simply took the seven classes suggested by Linnaeus and Wåhlin in 1756 (aromatic, fragrant, ambrosiacal, alliaceous, caprylic, repulsive, nauseating) and added two others, namely, ethereal odours (a class borrowed from Lorry) and empyreumatic or burnt odours (borrowed from Von Haller). Two of the systems concerned will be discussed here.

First comes Giessler's scheme[4] which, although quite unacceptable today, is of special interest because it is perhaps the first predominantly psycho-

[3] The chief odorous constituent of the skunk's secretion is *n*-butyl mercaptan.

[4] See p. 11.

logical scheme of modern times. He decided to ignore the objective — and especially the chemical — properties of odoriferous substances and to regard odours from the subjective standpoint. Basing his classification on the effect of odours on various parts and functions of the body he was led to distinguish two main classes, namely, those which have a predominantly *physical* effect *(i. e.,* also cause weeping, sneezing, *etc.)* and those which have a predominantly *psychic* effect. He also distinguished several sub-groups.

The next, published five years later, was a rather imprecise classification of the odours of drugs proposed by Lowe. Wishing to aid recognition of these, he employed such terms as "agreeable", "fragrant", "spicy", and "strong".

Matteotti, believing that the common speech of the people of Tuscany and, in particular, that of the inhabitants of Florence, included words capable of defining with great accuracy all degrees and qualities of odour, submitted a classification based on these words. No particular theory formed the groundwork of his "scale" which, he maintained, would be more useful despite its defects than would a more scientifically designed system. Like Zwaardemaker's scheme, Matteotti's has nine classes. They are: *profumi, aromi, fragranze, olezzi, odori, siti, tanfi, puzzi, and fetori.* Since the first four of these can be rendered in English, sweet smells, aromas, or fragrances and the last four, bad smells, stinks or stenches, it is difficult to understand the value — outside Italy at least — of such a classification. (The fifth term is translated simply "odours", or "smells".) The worth of Matteotti's paper, indeed, probably depends more on his interesting discussion of other aspects of olfaction than on his attempt at classification. Nevertheless, his paper was republished after an interval of forty years in the *Rivista Italiana Essenze Profumi, Piante Officinali, Olii Vegetali, Saponi* with annotations by Rovesti.

Those who have endeavoured to classify odours usually appear to begin by choosing certain broadly descriptive terms such as "flowery", "spicy" and so on. They then proceed to distribute all the odours known to them in such a way that each odour is ranged with those others judged, *by themselves,* to be most similar. Kennel, whose work was done a few years after that of the Italian investigator, adopted a somewhat different method. From eight of Zwaardemaker's groups, he chose thirty-nine odours. These he offered without any previous guidance to thirty-five persons (he recognized that some hundreds of subjects should be tested) belonging to "liberal" professions (professor, students, *etc.)* who were asked to divide the odours into groups. Kennel, in fact, did not wish to make a new classification. His aim, he says, was simply to apply the "method of majorities". This method, as he himself indicates, suffers from the disadvantage that the same value is attached to the opinions of all the subjects, whether they are competent or not. Most of the odorants he employed were mixtures such as coffee, mustard, rancid fat and so on but he also used pure compounds such as camphor, iodoform, and ether.

The conclusions which he reached are summarised as follows: in general, odours connot be definitely arranged in well-defined groups. Such groups as can be formed are "elastic". Many odours may, without risk, be placed in one group or another or may serve as transitions between one group and another. Of the classifications known to him, he regarded Zwaardemaker's as the best.

If not the best, it is at least the best known. Scarcely less familiar is Henning's system based on his "olfactory prism". General criticism of all the better known systems of classification has frequently been expressed. Henning's system, however, has also been the subject of specific and experimental criticism and it seems worth while to give a short account of these criticisms here.

He regarded all odours, it will be remembered, as forming a qualitative continuum represented by the surface of a triangular prism. At the six vertices of this prism are the six principal groups of odours which are analogous to the "Hauptfarben" of the colour pyramid and are fundamental reference points. At the corners of one triangular face of the prism lie the fragrant, ethereal, and putrid odours, and at the other end, respectively opposite these three, are the spicy, resinous and burnt odours. Psychologically simple and qualitatively intermediate odours lie along the edges and in the five faces but the interior of the prism is hollow.

One of the first critics was Ruzicka who easily showed that no reliance could be placed on the prism from the point of view of any relationship between chemical constitution and odour. If one wishes, he said, to arrange the odour relations in a geometrical form (as a preliminary step) one must be content with a sphere!

Another chemist who severely criticised Henning's claims was Otto Wallach. His comments elicited a very sharp reply but no real rebuttal.

Following Wallach came three American psychologists. The first of these, Dimmick, found that although some confirmation of the system could be obtained, cases in which the observers who performed the tests were in total disagreement with the German psychologist were far from infrequent. Dimmick, like other critics, realised the great lack of experimental evidence in Henning's writings but used the prism as a starting point and tested seventy-five of the odorants listed by him. (In later tests, Dimmick used a hundred odorants and thirty observers.)

Macdonald, whose work was done almost at the same time as that of Dimmick, used thirty-five odorants (chiefly definite chemical compounds) which were tested by three trained observers. The results of his study showed Henning's system to be logically and factually inadequate, the extent of correlation between the qualitative relationships as stated by Henning and as observed by Macdonald's subjects being only from 30% to 40%.

Finally, in a repetition of Macdonald's work in which stimuli found by Dimmick to give the greatest degree of constancy amongst his observers were used, Findley was led to conclude "that even when stimuli that show a relatively high degree of constancy in being placed in accordance with Henning's system are chosen, and when they are placed in the surface of the prism in the way to do the least violation to Henning's system, we are still unable to establish the validity of the system with any great degree of precision".

Henning himself was apparently unmoved by these criticisms which were directed against the account given in the first (1916) edition of his book. Indeed, in his second edition (1924) he claimed that his critics had misunderstood him. He emphasised, also, that his system, being a psychological construction, is concerned with subjective impressions and not with chemical constitution. Moreover, he even claimed that Zwaardemaker's classification, in its final form, was fundamentally in harmony with his own. But he adduced little or no further experimental evidence in support of his claims and, although Florence Hazzard later supported them, after testing six observers with fourteen odorants, her country-woman, Eleanor Gamble, affirmed that "not half the evidence to substantiate the prism has been found" and referred to the "statistical evidence which Henning notoriously fails to supply".

Many classifications have naturally been based on likes, dislikes and real or supposed similarities. Other bases, however, can be used. Thus Ohma, a Japanese worker, divided odours into sub-classes by determining how much, after they had been allowed to induce complete fatigue in the observer, they raised the threshold for other odours, a very important method which has since been developed by others. He conducted his experiments in a *camera inodorata* (capacity 64 litres) with odorants such as borneol, camphor, citral, safrole and pinene and, to provide a "physical basis" for his procedure, he also classified by number of atoms and atomic groups in the odorants, by their indices of refraction and by their diamagnetism. His fatigue experiments led him to distinguish three sub-classes, namely, the almond odour, the camphor and spice odours, and the citral odours.

The Italian classification discussed above was claimed to be "natural". It was followed after a lapse of almost twenty years by another Italian system said to be "fundamental". *(Una classificazione fondamentale degli odori.)* Azaretti, the author of this arrangement, suggested that odours should be classified according to the primitive activities which they stimulate in the human being. He therefore divided odours into five main classes, each of which had several (two, three, or four) subclasses. The main classes were: (1) the appetizing odours, those which indicate the presence of edible matter and stimulate the appetite; (2) the balsamic odours, those which indicate the presence of good air or of substances which accompany such air and which stimulate vigorous respiration; (3) the erotic odours; (4) the loathesome odours, those which are emitted by substances chemically or bacteriologically dangerous when introduced into the organism or by materials which accompany such substances; (5) the irritating odours, those belonging to substances having a violent, destructive action on the mucous membranes, especially the olfactory mucous membrane. The presence of these substances is directly dangerous to the organism.

The systems so far mentioned are all entirely qualitative. But quantitative, or at least semi-quantitative systems have also been devised. Thus, in 1927, Crocker and Henderson, while attempting to discover "the elements of sensation which make up all odors", developed "a semi-quantitative evaluation of odors making possible the accurate description of any odor by the simple device of a four-digit number. As examples of this code the odor of a rose may be stated as 6423, the odor of acetic acid as 3803, and that of freshly roasted coffee as 7683. Odor descriptions with this system are even simpler than telephone numbers, for the "ring" or "party" letters do not yet appear to be needed and the nose is the only "exchange"." Crocker and Henderson took Henning's system as their starting point but, like the other American workers, they were led to question his fundamental types. Eventually, they came to consider that there are only four elementary odour sensations which are "the principal and perhaps the only units which make up all the odors we perceive". These four are: the fragrant or sweet, the acid or sour, the burnt or empyreumatic, and the caprylic or oenanthic. All odours are regarded, then, as being compounded from these elementary sensations, mixed or combined in varying "amounts", the digits used indicating these "amounts". The number assigned to Tonquin musk, for example, is 8476. This shows that it is regarded as having eight units of fragrance, four of acidity, seven of burnt odour, and six of caprylic odour.

This is undoubtedly an ingenious and possibly useful scheme which combines both the qualitative and the quantitative aspects of the subject. But it seems rather unlikely that any but highly trained persons could distinguish the several components and their strengths in other than a few particular cases. Moreover, no convincing evidence is given for the assumption that only four elementary olfactory sensations exist or that these are the sensations chosen.

Crocker and his co-workers, however, have continued to maintain that their system is sound and they have received considerable support from other workers including Dorough and his colleagues who classified twenty-two octanols by the method and Plank who, in 1944, thought it the best system to date. And an American firm has offered for sale sets of the Crocker odorants in sniff bottles. In 1947, Crocker stated that the system is based on the following three assumptions: (1) that the human nose is provided with four and only four kinds of odor nerves; (2) that every odoriferous substance, sniffed in adequate amount, stimulates all four kinds of these nerves simultaneously to the extent characteristic of the substance and of its concentration, and (3) that the odor sensation chord thus created by the excitation of the nerve endings is capable of producing a distinctive odor impression. These assumptions, he added, have not been rigidly proved but the system, during twenty years or more, has "worked". With Dillon, Crocker tested over five hundred substances used in perfumery and later published an "odor directory" in which about two hundred and fifty aromatic chemicals and more than a hundred natural odorants were classified both alphabetically and in numerical sequence in accordance with the system. Further, he rejected Sagarin's assertion concerning lack of reproducibility, pointing out that the system had stood the test of time and experience. A trial with four experienced observers and four novices had shown, he affirmed, that they exhibited only slight deviations from the established values. However, when Ross and Harriman tested the system in 1949 using thirty-two odorants and fifteen untrained subjects each with a good sense of smell, they found wide variability and concluded that the system is inadequate.

Another fairly recent classification, that of Elsberg and his colleagues, is also quantitative though it appears to lack the qualitative basis which all systems would be expected to possess. Elsberg devised an olfactory "test bottle" provided with nosepieces by means of which an exactly measured volume of odorous air could be delivered, by injection, to the olfactory mucous membrane. (An undesirable feature of the apparatus is that it includes a stopper and tubing made of rubber.) He came to the conclusion that, in normal individuals, the most important factors in perception — other than the specific odour quality of the substances concerned — are the volume and pressure of the current which impinges on the membrane. His "new principle for the classification of odors" is "based upon the olfactory coefficient of each odor, that is upon the measurable minimum volume and pressure that is necessary to produce an olfactory sensation for the identification of the odor". He claimed, also, that the coefficient (for substances liquid at ordinary temperature, at least) varies directly with the boiling point. Classification, then, according to Elsberg, consists in arranging the odorants "in order from those substances which required the smallest volume and the lowest pressure to those which required the largest volume-pressure for their identification". The coefficients are given by volume (cubic centimetres) only. Presumably the air ejected from the bottle is assumed to be saturated with the vapour of the odorant though this is not clear and, in any case, nothing is said about the weight of substance actually conveyed to the membrane.

Obviously, in so far as this is a valid system, it is a system of odorous strengths, not a classification of qualities. Elsberg admits that the arrangement may seem incongruous but he maintains that the incongruity is more apparent than real. Nevertheless, it is surprising to find, for example, that the odour of ground coffee is placed in an intermediate position between the odour of oil of sassafras and that of oil of anise! It is also surprising that few odorants were tested, that some of these were natural oils containing a great variety of compounds, and that the laws governing olfaction are believed to be the same as those governing vision.

Since the Elsberg scheme was devised, there have been a few other attempts at classification. One was made in 1937 by the Japanese psychologists Kido and Yoshida but, apparently, they asked their subjects only to report on the pleasantness or unpleasantness of the sixty-six odorants used.

Unusual interest attaches to a test made by Pastier in 1939 of the arrangement proposed by Heyninx in 1919, an arrangement in which there are seven fundamental odours, namely, burnt, ethereal,foetid, rotten, sharp, spicy, vanilla. Pastier tested eighteen normal persons and thirty-five mental patients with these odours and came to the conclusion that Heyninx's classification is physiologically good but psychologically inadequate. It is worthy of note that no other experimental examination of this system seems to have been made.

During the second world war and afterwards, the distinguished French investigator Le Magnen studied the possibility of making a classification by the "co-saturation" method, that is, by inducing fatigue for one odour and then offering others. Those others not then perceived or not readily perceived are regarded as being in the same class. In other words, if fatigue for odour A also induces fatigue for odour B, then A and B are in the same class. (This was the procedure adopted by Ohma many years earlier.) Le Magnen determined thresholds immediately and after ten minutes of adaptation (fatigue). In his earlier experiments he used a variety of odorants (*e. g.*, butyric acid, terpineol, vanillin) and in later experiments, almond, camphor and musks. Since he found that the odours of the macrocyclic musks and those of the nitro-musks, although apparently very similar, do not affect each other, he pointed out that subjective judgment is not always correct.

Just after the war (in 1946) Hsu applied modern factorial technique in an attempt to single out the primary factors of olfaction, asking 182 subjects to rate twenty-one odorants (chiefly pure compounds having little or no trigeminal effects) on a pleasant-unpleasant scale. His results, however, were inconclusive and disappointing.

Equally, or more disappointing is the very elaborate scheme proposed by Voge and Sinclair some years later. This seems to be based partly on the decimal classification and partly on the Crocker-Henderson system but the authors' description of it is rather incomprehensible and, in any case, it is difficult to see what purpose such a classification could serve.

Perhaps the most recent classification (other than that of Amoore which will be discussed in the section on adsorption) is Moncrieff's. He proposes to group odours on the basis of the rapidity with which or the extent to which the corresponding odorants (foods such as cheese, coffee and fish included) are adsorbed on materials such as activated alumina, activated carbon and silica gel. This proposal is related to the assumption that, in the olfactory system, there are several kinds of receptors having different adsorptive powers and that the differences may account for differences in the odour qualities perceived. Moncrieff's experiments with the adsorbents mentioned and with the respiratory and olfactory passages in intact sheep heads showed, as those of several earlier workers had already done, that some, if not all, odorants are very readily adsorbed on the surfaces of inorganic and organic materials. It seems highly unlikely, however, that this property alone is related to the quality or the intensity of the odours and it is unfortunate that he did not examine the ease (or difficulty) with which *odourless* substances of comparable molecular weight are adsorbed on the same materials. *It may be added here that failure to test such substances for comparison when experiments with odorants are being made often completely deprives the experiments of their value.*

In the present state of knowledge, it seems unlikely that a fully satisfactory classification of odours can be made. If, as is probable, the olfactory region contains several types of receptors having differential sensitivity, it follows, *perhaps,* that there are as many primary odours and that the number of secondary or "mixed" odours which could be perceived by stimulating pairs or larger numbers of different receptors could be calculated. Enthusiastic students of information theory have already done something like this. But account must be taken of the fact that the quality of some odours changes very greatly with change in concentration and also of the fact, so well established by French workers, that a single substance *(e. g.,* exaltolide) may have no odour or several odours according to the sex, age and (in females at least) physiological condition of the normal healthy percipient.

The difficulties of classification are particularly well known to perfumers, of course, and they may often invent systems of their own for private use. It is not surprising, then, that learning the perfumer's art involves very prolonged training and that, as one of the best of them, Carles, has recently said "Sans enthousiasme, on ne sera jamais parfumeur". On the other hand, as Le Gros Clark has pointed out, "... it may seem surprising that an elementary sense such as that of smell, depending on stimuli initiated by simple chemical substances which often have an exceedingly simple molecular structure, and mediated by nervous pathways which in certain respects are a good deal less complicated and tortuous than those of other senses, is still so little understood in anatomical and physiological terms. But this only serves to emphasise how incredibly complex is any problem whose basis is neurological. The history of neurological research makes it abundantly clear that investigation into the functions of the nervous system have always been, and will certainly continue to be, a slow and arduous process." It may be added, in conclusion, that the experiments with the co-saturation method are being continued by Cheesman in Tasmania with promising results and that, in the U. S. A., Schutz has very recently proposed a classification system similar to that of Crocker and Henderson but considerably more refined.

(During recent years, some success has been attained in describing and classifying odours by the "polarity profile" method which is said to yield reproducible results. Eyferth, who has applied this method, finds that the "odour space" has "seven dimensions".)

The Brave Smell of a Stone

> One had to run over the whole gamut of odours, some so faint that they embraced the nostril with a fairy kiss, others bluntly gross, of the knock-you-down order; some sweet, with a dreadful sourness; some bitter with a smack of rancid hair-oil. There were fine manly smells of the pig sty and the open drain, and these prided themselves on being all they seemed to be; but there were also feminine odours, masquerading as you knew not what, in which penny whiffs, vials of balm and opoponax, seemed to have become tainted, vaguely, with the residue of the slop-pail.
>
> Their house had a strange, delicious smell, so unlike anything I smelt anywhere else that it used to fill my eyes with tears of mysterious pleasure. I know now that this was the odour of cigars, tobacco being a species of incense tabooed at home on the highest religious grounds. *E. Gosse*

Since it seems certain that volatile material only can be smelled, an explanation must be found for the fact that many non-volatile substances undoubtedly appear to emit odour. It is said, for example, that some people can smell sugar in tea or salt in soup although these solids are non-volatile and, in the undissolved state, odourless. (Andres, in response to the question: can sugar smell?, showed that it readily absorbs odours from sacks and other materials in or near which it may be stored and also that white sugar (cane or beet) may give off odour after storage for several months in tins or glass jars. More surprising still is the odour attributed to such materials as rocks, clay and metals (*e. g.*, coins, turnings, filings)[1]. It cannot be

[1] As early as 1798, the "odours" of metals were discussed by Fourcroy. He maintained that several metals, especially copper, iron, zinc, and tin have very strong "odours" and that the "odour" of copper, for example, is not due to some "principle" exhaled by the metal but is due to molecules of copper which are carried into the nose.

supposed that particles of these substances penetrate into the olfactory cleft and so come into contact with the olfactory mucous membrane. Whence then comes the odour? [2]

This question was examined, in 1898, by Ayrton in co-operation with the ladies of his family (Mrs. Ayrton found the 'odour' of aluminium particularly offensive), an account of the investigations being given in his presidential address to Section A of the British Association at Bristol in that year. The metals studied included brass, aluminium, bronze, copper, gold, iron, phosphor-bronze, silver, tin, steel and zinc. During the experiments, it soon became clear that the odours perceived were not those of the metals themselves but of impurities and substances produced by the action of moisture (in the breath, for example) and of perspiration. It was also shown that surfaces of metal and glass tenaciously adsorb odorous substances as later experiments by other workers have abundantly demonstrated.

It seems that more than thirty-five years passed before another scientific study of the metallic odour was made. (Some experiments by Gruhn were reported in 1907 but their value is somewhat doubtful.) This investigation was carried out by Raub at the Institute for the Study of Noble Metals at Schwäbisch-Gmund, Germany. In the years preceding 1934 — the year in which the account of this study was published — numerous enquiries about the unpleasant odour (and taste) of metals had been received at the Institute and specimens sent with some of the enquiries invariably had a "bad smell and taste". Raub devoted most of his attention to silver but he also examined nickel, chromium, zinc, iron, aluminium, gold, platinum and several alloys including brass and bronze. Since not all specimes of silver had an odour and since the strength of the odour, where it occurred, varied from specimen to specimen, he at once concluded that the cause must be impurity in the metal or foreign matter on its surface. Selenium-silver and tellurium-silver alloys containing up to 0.5% of selenium or tellurium (compounds of these two elements have powerful unpleasant odours) however, yielded no odour when exposed to the conditions to which silver articles are normally exposed. Accordingly, suspicion fell on material adsorbed on the surface. A study of the whole course of silver manufacture failed to reveal contamination with odour-producing material. Odorous specimens were next examined. It was found that the odour was difficult to remove. It could not be eliminated by boiling with soda solution or by rubbing with polishing powder. On the other hand, hot, dilute hydrochloric acid (but not other acids), nascent hydrogen, and heat (temperature 250—300 °C) destroyed it more or less rapidly. (Gentle heating, in the hands for example, and treatment with cold dilute hydrochloric acid increased it at first.) Moreover, the

[2] According to Hiorns, the vapour of arsenic has a peculiar garlic-like odour and that of cadmium has a disagreeable odour.

hydrogen used acquired an unpleasant odour as did the distillate obtained when the hydrochloric acid used was distilled with steam. These results suggested that the substances responsible for the odour might be sulphur compounds (mercaptans, thio-ethers), especially as such compounds are present in vegetables (onions, for example) and in human perspiration. Raub therefore exposed strips of foil to the action of onion, leek, and garlic and to the steam-distillate from the water in which these vegetables had been cooked. In these circumstances, silver always acquired a strong unpleasant odour which it did not lose during weeks of storage in pure moving air. Mercaptans and thio-ethers had the same effect as the vegetables and the distillates. Copper, brass, bronze and copper-silver alloy were also rendered strongly odorous by these treatments but the other metals were only slightly or very slightly affected.

Another aspect of the mineral odour was examined, ten years after Ayrton's work was done, by Jules and Auguste Piccard. Their attention had been drawn to the peculiar odour produced when pebbles, especially the white pebbles carried down from the Alps by glaciers and streams, were struck against each other. This odour, said to be sulphurous or like that of phosphorus, they likened to that of burned horn or overheated wool. For an explanation of the cause of the odour they turned to the geologists, the physicists and the chemists but these, though they knew the odour in question well, could not explain it. Accordingly, the Piccards themselves conducted a series of experiments with many very hard minerals such as quartz, rock crystal, agate, flint, granite, emery and corundum. The conclusion reached as a result of these experiments was that, "beyond any doubt, the odour emitted when the stones are struck is due to the volatilisation and the decomposition by heat of traces of aliphatic nitrogenous organic matter which quasi-infallibly occurs on the surface of these minerals". They found, in the first place, that no odour was produced when pure quartz pebbles were struck together under conditions in which all contact with external organic matter was rendered impossible. This matter may be derived from atmospheric dust, lichens, moss or algae, or from earth or mud on which the stones have rested. But chiefly it is derived from the *hands* which have held the stones, no matter for how short a time. A minute quantity of organic matter from the epidermis or the cutaneous secretion suffices to produce the odour when the blow is struck. The reaction involved, say the Piccards, has a 'truly unbelievable' sensitivity and tenacity. Consequently, if it is to be prevented, energetic steps must be taken. It is not always enough to wash the pebbles with caustic soda or to brush them with a metal brush. They must be heated to a dull red heat in a muffle furnace or boiled for half-an-hour with a mixture of chromic and sulphuric acid. And, above all, they must on no account be allowed to come into contact with the fingers or the palm of the hand. Gloves of newly washed linen should be worn.

With respect to this 'mineral' odour, metals behave like the pebbles. A stretched platinum wire which has been fingered emits the characteristic odour of struck pebbles when slowly heated electrically to about 200 to 300 °C. When dull red heat is reached, this odour disappears but it can be reproduced by allowing the wire to cool, touching it with the fingers, and reheating. Gold, platinum sponge, and dental amalgam can be shown to adsorb organic matter from the skin in the same way. In order to prove that the temperature attained by striking pebbles suffices for the decomposition of organic matter, the Piccards purified pebbles with chromic acid, rubbed them with a mixture of arsenic oxide and sodium acetate or carbon, and struck them together. When this was done, the odour of cacodyl or garlic was observed. The reactions involved require a temperature of at least 300 °C. Similarly, the odours of caramel and sulphurous acid were observed when sugar and sulphur respectively were used in place of the arsenic oxide, sodium acetate, and carbon.

Confirmation of the Piccard's view was soon forthcoming. In the summer of 1908, Fleissner had occasion to prepare an artificial stone at a temperature of 1300 °C. Pieces of this material, when freshly prepared, emitted no odour when struck against each other. But when the hand was passed, but once only, over the surface newly exposed by breaking before the striking took place, the 'odour of striking' was distinctly perceived. The newly-made material, of course, contained no organic matter because of the high temperature employed in the manufacture.

So-called mineral odours are not confined to metals and hard substances such as quartz. Cornu found that almost all mineral 'hydro-gels', not only, as had previously been supposed, the aluminium silicates, emit the 'odour of clay' and also other odours when moist. It can easily be shown, he maintained, that all these odours are due to adsorbed substances and he rejected the view, advanced by Semiatschenski, that very minute solid particles of the minerals are responsible.

In 1913, the observations of the Piccards were again confirmed and also extended by Schwartz in Darmstadt. He pointed out that although the odour produced when stones are struck together is sometimes due to bituminous inclusions, this is by no means always the case. Many minerals and artificial silicates and some non-silicates, magnesite for example, yield the odour. (Amongst the artificial silicates are glass, porcelain, and earthenware.) Silicates which do not yield it include serpentine and asbestos. Schwartz examined the collection of minerals in the technical college of Darmstadt and found that all the silicates and all varieties of quartz having hardness greater than 3.5—4.0 yielded the odour which usually persisted for periods ranging from 30 seconds to three minutes. The intensity of the odour, he concluded, seems to be a function of the hardness and heat-conducting power of the mineral, greatest intensity accompanying hardness between 5.5 and 7.0. He conducted experiments with natural silicates and glass rods and found that the adsorbed material responsible for the 'frictional' odour could not be removed with organic solvents. This aim was achieved, however, by heating for five or ten minutes at 300 °C or 260 °C. Moreover, he found that glass rods which had been rubbed together so long that the surface layers were converted to powder and so removed likewise yielded no frictional odour although they did so before the removal of these layers. Similarly, surfaces of silicates and other rocks, freshly exposed by breaking with a hammer did not yield any odour. On being exposed for several weeks, however, these surfaces acquired the power to yield odour. Finally, Schwartz sought to throw light on the nature of the material responsible for the odour. He rendered some silicates odourless by heating to red heat and then allowed them to come into contact with various pure solid or liquid organic and inorganic substances. The inorganic compounds had no effect. Many carbohydrates and nitrogenous organic compounds had the effect of conferring frictional odours not similar to that normally observed. When proteins or protein-like substances were used, however, the resemblance was close. It was also found that the odour could be developed after merely touching the surface of the stone with the finger. Evidence was thus provided for the view that the matter responsible for the odour is a protein-like material adsorbed by the minerals from the air or elsewhere and decomposed by the heat generated by friction when appropriate conditions of hardness and heat-conductivity exist.

These conclusions were strengthened, in part, by Professor Johnsen of Kiel who likewise observed that the odour of 'singed horn' was absent when freshly exposed surfaces of flint were rubbed together. When the surfaces were rubbed between the hands "as one rubs a piece of soap when washing" however, subsequent friction produced the odour.

Commenting on Johnsen's observations, Smekal of Vienna pointed out that, in the Alps, another mineral odour, quite distinct from the singed horn odour was frequently encountered. This 'burnt' or 'sulphurous' odour, it seems, is well known to mountaineers and to those who live at high altitudes where landslides and avalanches provide the cause of the friction. Smekal remarked that this odour is independent of the nature of the rocks concerned but he also suggested that, in some cases at least, traces of pyrites may be the cause. Another suggestion is that lichens, growing in crevices of the rocks may be responsible. Whatever the cause, the odour is certainly encountered and is not merely an invention of hunters and tourists. The people know it well. Evidence of this is provided, for example, by the popular account, dating from 1768, of the sulphurous odour which persisted for several days after a certain notable landslide.

The view that there are several mineral odours was also expressed by Brandt who, on the basis of his own experience, concluded that there are at least four, namely, the 'dust' odour, the odour of singeing, the odour of putrefaction and the 'chemical' odour. Solid particles of dust, he believed, may come into contact with the olfactory mucous membrane, so giving rise to the 'odour of clay'. The odour of singeing or of singed horn is that described by the Piccards, Schwartz, and Johnsen. It closely resembles the odour of burning horses' hooves. Of the four odours, it is the only one solely derived from surfaces *(i. e.,* surfaces which are not freshly exposed), the others being emitted by the whole mass. The odour of putrefaction is derived from bituminous material. It is *more* pronounced when emitted by freshly exposed surfaces than when derived from older surfaces. Quite characteristic but difficult to define is the 'chemical' odour. Brandt observed it in an underground chalk pit whilst the ground above was undergoing bombardment by artillery. He suggested that it might be derived from inclusions of pyrites. It was his opinion that the search for an explanation of these odours might be of great value to geophysical, petrographical, and physiological science. And the interesting discussions of the subject which were published about this time by Heller and Schelenz certainly do not in any way contradict this view. Heller, indeed, who writes very lucidly and forcibly, insists that all so-called mineral odours, such as those of the 'violet stones' or 'wonder stones' which emit distinctly pleasant odours when gently rubbed or when carried in the pocket, emanate either from included or adsorbed organic matter or, in a few instances, from volatile oxide or sulphide; and Schelenz, in a rather long-winded and semi-philosophical fashion, merely confirms this conclusion.

The most recent communications on the 'mineral odour' are probably those of Halla and his collaborators, of Oakley, of Lucas and of Haler. In their letter of 3 February 1944 from the Institutes of Physical Chemistry and Mineralogy of the Vienna Technical College, Halla and Schmölzer once again prove that the 'burnt horn' odour is due to thermal decomposition of material absorbed on the surfaces concerned. They consider that this material consists of protein from soil bacteria. Like earlier investigators, they find that freshly exposed surfaces (of granite and quartz) yield no odour. Further, they find that the adsorbed material, and with it the power to yield odour, can be destroyed by treatment with chromic acid.

Twelve years later, Halla, on this occasion with Van Tassel in Brussels, showed by direct experiment that when powdered emery, carborundum, felspar, commercial titanium oxide, and also pure specimens of artificial corundum, quartz and calcite were rubbed from five to thirty times between the palms of the hands and then dropped into a red hot platinum dish the 'burnt horn' odour was produced, the intensity of this odour increasing with the hardness of the substance and the frequency of rubbing. Moreover, the surprisingly large amounts of kreatin removed from the skin by the rubbing was revealed by the brown colour which appeared on ignition of the pure white powders. Control specimens yielded neither odour nor colour.

Oakley's remarks at a meeting of the Geological Association in 1945 were made during a discussion of a paper by Hawkes on fragments of jet coal in the chalk of Kent. He reported no experimental work but merely referred to the smell of burning horn emitted by flint when struck.

The experiments of Lucas with three specimens of 'stinking chalk' proved that these contained the ammonium salt of a volatile acid, probably organic, containing phosphorus but no sulphur.

Finally, Haler returned to the question of the odour of metals or rather of iron and its salts. Reading an article on ferrotherapy, a subject in which he was extremely interested, he encountered a report on biological specimens which smelled of iron and of ferrous sulphate. This led him to submit iron powder and six iron salts to olfactory tests. All these were completely odourless to himself and to his six observers and he felt obliged to protest against statements that these inorganic materials can be smelled. He pointed out that iron salts have a characteristic metallic taste and the editor of the *Lancet,* in which Haler's communication appeared, suggested that "anything with a strong taste may often be assumed (wrongly perhaps) to have a strong smell". The editorial comment, however, adds that "it seems unlikely that this is the whole explanation".

The argillaceous odour produced in dry clays and soils by moistening with water has been re-examined very recently (1964) by Bear and Thomas who refer to the fact that drought-stricken cattle are said to respond in a restless manner to this "smell of rain" and who tell that, near Kanauj in India, the production and concentration of earth perfume *(matti ka attar)* has formed the basis of a small perfume industry. Baken clay discs, exposed to the sun during the hot summer months are boiled with water which is then steam-distilled, the vapours containing the odorous material being absorbed in sandalwood oil. Many types of rock and mineral aggregate, whether porous or not, evolve the odour but silica and silicates are outstanding in their capacity to yield it. Bear and Thomas propose the name 'petrichor' for the odour. By steam distillation and the use of solvents they extracted a small quantity of the oil responsible for it. This oil was found to be very complex in nature. It appears to contain a nitrophenolic compound and a basic material having an extremely powerful odour resembling those of quinoline and pyridine.

It may be added here that certain alloys, first used in recent years, yield characteristic odours when fractured or machined. It seems very probable that these odours, which are also observed in slags, are derived from traces of arsenic, phosphorus and sulphur compounds present in the metals.

Physiological Mystery

Shall any gazer see with mortal eyes?
Or any searcher know by mortal mind?
Veil after veil will lift: buth there must be
Veil upon veil behind. *Arnold*

Atmest du nicht mit mir die süssen Düfte?
O, wie so hold berauschen sie den Sinn.
Geheimnisvoll sie wehen durch die Lüfte,
Fraglos geb' ihrem Zauber ich mich hin. *Wagner*

We may note, in the fields of olfaction and gustation, not that the theoretical position is insecure, but simply that no theoretical position exists.
Rawdon-Smith

The theories of olfaction have certainly been incomplete and rather unsatisfactory. Some of them are largely speculative. Others are philosophical, dealing with *perception* rather than *sensation*. But it is no longer true to say that "no theoretical position exists". In this chapter, some of the theories will be reviewed and remarks will also be made on various other interesting aspects of olfaction.

Anatomy

Of all the sensory functions of the human body, that of olfactory discrimination seems the most mysterious. *Le Gros Clark*

The sense of olfaction still taunts the physiologist with his lack of knowledge of even its most basic aspects. *Griffin*

Descriptions of the olfactory systems of human beings and animals have frequently been published. They need not be repeated here but perhaps special attention should be directed to the papers of Allison (1953), Brodal (1947), Engström and Bloom (1953) Le Gros Clark and co-workers (1946, 1947, 1957), Ferrari and Giuliani (1957); to Pocock's (1916) account of the scent glands of animals; to Bronstein's (1960) studies of the localization of enzymes (choline esterase, alkaline phosphatase, *etc.)*, glycogen, protein, SH-groups and ribonucleic acid in the olfactory cells of the rabbit and to the finding by Miline and co-workers (1963) that, in rats and mice, prolonged exposure to olfactory stimulation with a (to man) malodorous *iso*-nitrile causes changes in the internal nucleus of the habenula and atrophy of the pineocytes of the pineal gland.

Pigment

> We might call smell the sense of memory, because it recalls to us more directly than any other the specific impression of an event or scene even from the most distant past. *Schopenhauer*

> Smells are surer than sounds or sights
> To make your heart-strings crack —
> They start those awful voices o' nights
> That whisper, "Old man, come back".
> That must be why the big things pass
> And the little things remain
> Like the smell of the wattle by Lichtenberg
> Riding in, in the rain. *Kipling*

In man and other mammals *(e. g.,* cat, dog, fox, rabbit, sheep) the olfactory epithelium has a yellow-brown colour which is absent from the surrounding tissues. Nearly a century ago, Ogle suggested that there is a correlation between the proportion of the pigment concerned *(i. e.,* the intensity of the colour) and olfactory acuity. He claimed that dark-skinned human beings have greater acuity than have white-skinned and he believed that albinos should prove to be anosmic. Farmers, he said, dislike albino animals and do not breed them because, being partly or wholly anosmic, they fail to recognize the odours of poisonous plants and suffer accordingly. And he referred to the belief that dark materials *(e. g.,* fabrics, plant tissues) are said to absorb odorous substances more readily than do light.

Since Ogle's day, the question has been studied on several occasions and, in recent years, the chemistry of the pigment has been examined by Gerebtzoff and his co-workers and, especially, by Jackson. From these investigations it appears that the pigment consists chiefly of lipids or is adsorbed on phospholipids or carotenoids of the lecithin type. But as yet there is no agreement concerning the correlation claimed by Ogle. His opinion has been supported by Niccolini (1940), Onagawa (1957), Briggs and Duncan (1961) and Wright (1964) and rejected by Le Magnen (1951), Gerebtzoff and coworkers (1957—58), Jackson (1960) and Moulton (1960). Niccolini based his support chiefly on theoretical considerations *(e. g.,* that the pigment can have no light-receiving role) but Onagawa claimed that the sensitivity to olfactory stimulation is proportional to the depth of colour of the olfactory mucous membrane and, correspondingly, that the value of the electrophysiological action potential is also proportional to the degree of pigmentation. Depth of colour and degree of response, he affirmed, are definitely related. In the view of Wright, who has devoted much attention to it, the pigment may supply the energy necessary for olfactory excitation — it may have a "triggering" action somewhat similar, perhaps, to the catalytic action that Onagawa attributes to the secretion of the membrane. Briggs and Duncan believe that the coloured protein-bound carotenoids may be "molecular receptors of odours", having found that of 56 patients suffering from uncomplicated anosmia, fifty recovered their sense of smell partly or wholly after being given intramuscular injections of vitamin-*A*.

Vitamin-*A* deficiency, on the other hand, was not found by Le Magnen to be accompanied by much loss of olfactory acuity in white rats. Since albinos and some species of animals that have no olfactory pigment nevertheless seen to have normal olfactory acuity, it appears likely that, if the pigment plays any part in the mechanism of olfaction, it is seldom, if ever, an important part. It has been suggested, however, (by Campbell, 1956), that the carotenoids of the olfactory hair cells serve as quasi-metallic conductors in which the standing potential is suddenly altered on contact with an odorous substance. The extreme sensitivity of the olfactory receptors, indeed, demands for its explanation, the production of a "mega-event" (*i. e.*, the generation of a nerve impulse) from a "micro-event" involving a small number of quanta. (In a very recent paper (Feb. 1968) by Rosenberg and co-workers, an olfactory transduction theory involving the formula of a pigment-odorant complex is advanced.)

The Bloody Route

From time to time it has been suggested, chiefly by Italian workers, that certain odorous substances, injected intravenously, reach the olfactory nerve endings without entering the respiratory system and produce sensations of smell similar in quality to those experienced when the substances are inhaled in the ordinary way. The effect was said by Bednar and Langfelder in 1929 to be produced even when no inhalation of the substances is possible and the time taken to cause the sensation is that calculated for the transport of the material from the site of injection to the nerve endings. These investigators also stated that no perception occurred in complete anosmia and that rapid fatigue of the olfactory organ took place.

A few years later, Dussik and Kauders (1933) gave intravenous injections of odorous substances (*e. g.*, camphor, menthol) when testing for injury to the brain and for recovery from the injury. They claimed that the precision, flexibility and independence of the momentary condition of the patient conferred on this method advantages making it superior to the ordinary procedure in which the substances are inhaled. (In 1948, another aspect of this "haematogenic" olfaction was examined by Le Magnen. He found that intramuscular injection of odorous substances such as amyl salicylate, camphor, ether and eucalyptol lowered, often very greatly, the olfactory threshold for these materials when subsequently inhaled.)

Semeria, whose psycho-physiological work was done more than twenty years later, appears simply to have assumed that odours can be perceived directly by way of the blood stream. He administered peppermint intravenously to 83 normal persons and determined their thresholds of olfactory perception. He observed that fatigue occurred in many cases. In some other tests, he used amyl acetate, acetic acid, cloves, pyridine and vanilla as odorants. Other persons that he examined had abnormal olfactory acuity.

Five years later, his countryman Antonelli conducted electro-physiological tests on cats after giving them injections of citral, jasmin or peppermint. He pointed out that, if haematogenic stimulation actually occurs, it might be by direct action on the cerebral olfactory centres or by action on the olfactory receptors *via* the lungs. His conclusion was that direct stimulation of the receptors occurs.

To several other investigators, however, this conclusion has seemed unfounded. Van Dishoek and Versteeg, for example, were unable to produce haematogenic olfactory sensations with camphor, ether and lavender. (In some of their tests, they obstructed the nasal passages by means of a rubber bulb filled with water.) In their opinion, it is very improbable that haematogenic stimulation of cerebral olfactory centres occurs. Where (as with ether) the odour is perceived, it is derived from expired air rendered odorous by the blood in the lungs. It is also possible, they believe, that the saliva could be the source of the odorous material.

Vaglini, who administered peppermint intravenously after introducing a fine spray of acetylcholine, hyaluronidase or adrenalin into the nose also concluded that the so-called haematogenic action is actually a peripheral phenomenon especially as olfactory sensations were often absent during apnoea. And Bartalena, who studied the electro-encephalograms of persons to whom isotonic solutions of peppermint were administered intravenously likewise decided that, if odour is derived from substances so given it is merely because they escape into the pulmonary air which then passes over the olfactory receptors during exhalation.

Lastly, we may note the finding by Guillot that the threshold for *iso*-valeric acid is *not* affected by injecting an antibody specific for the acid. This also gives no support to the view that odours can be perceived haematogenically.

It is obvious that further work on this subject is needed and that, at present, the evidence for the occurrence of haematogenic olfaction is not very convincing. Indeed the very recent studies of Güttich (1965) seem to prove beyond doubt that "haematogenic" smelling is simply gustatory, *i. e.*, the odorant reaches the olfactory region in air from the mouth.

Chemistry

It cannot be upheld indefinitely that the alleged relationships between molecular structure and odour have ever led to the creation of a new type of odoriferous substance. These relationships are summary, of very limited scope, and the parrotry which prevails concerning them endows them with absolutely no new significance; it merely suggests minds that are unfamiliar with the history of the perfume industry. *Y.-R. Naves*

When the structure of an unknown organic compound is drawn with all substituents in their correct positions and with complete indications of configurational details, we have at the same time irrevocably and fully defined all the properties, chemical, physical and physiological, including the odour which the compound will have as soon as it has been synthesized. In other words, we need not consider the question whether a relationship exists between structure and odour because this relationship must be complete and unabiguous. The only question which interests us is whether it is simple enough to be detectable with our limited intellectual and technical means.

M. G. J. Beets

Although it seems impossible that there should be no relationship between the odour of chemical compounds and their constitutions it has to be admitted that the difficulties in the way of uncovering the connection are very great. Consider some awkward points that call for explanation. Haldane has affirmed that carbon monoxide, usually said to be odourless, actually has an odour like that of garlic and he suggests that the gases of the atmosphere may have odours which could perhaps be perceived if they were present in significantly high concentration and not breathed continuously. Jellinek has pointed out that odourless vehicles used in perfumes have a profound effect on the odour, some odorants smelling much stronger in water than in mineral oil and *vice versa*. It is well known, too, but none the less odd, that many odorants have one odour when concentrated but quite a different odour when dilute. (Guillot explains this by supposing that the concentrated substance produces immediate fatigue. He purified vanillin fractionally by sublimation and dissolved the fractions in alcohol. No difference in odour could be detected between the first and last fractions. After a certain degree of purity had been attained, further purification had no qualitative or quantitative effect on the well defined odour.) Robert Boyle's "eminent professor of mathematicks", on the other hand, could give no explanation of the curious fact that the large dunghill in Lincoln's-Inn Fields smelled strongly of musk when at a distance but had a stink proper to such a heap of excrements when close at hand. And even more puzzling is the fact that many odorants (*e. g.*, hydroxycitronellal, phenylethyl alcohol, vanillin) lose all or part of their odour in proportion as they are more and more highly purified. Thus the pleasant odour of commercial geraniol is appreciably altered by purification although the material separated in the process does not have the odour which the geraniol has lost. Again, substances such as acetylene and arsine, generally said to have very unpleasant odours, are believed to be odourless when pure. Even skatole completely loses its objectionable smell when subjected to zone refining. But impurities, if actually responsible for these odours, could presumably be purified in turn and so deprived of *their* odour and we might be tempted to ask: are absolutely pure substances always absolutely odourless? If perchance no absolutely pure substance has an odour, whence does the odour come?

Whatever the answer to this question, it is very obvious that a substance regarded as pure from the chemical point of view may be very far from pure in the olfactory sense. However, since it is now possible by modern methods (*e. g.*, the various forms of chromatography) to attain very high degrees of chemical purity, the attainment of very great olfactive purity should also be possible.

In order to throw light on this problem, Guillot studied some *isosteric* substances which, when present as impurities, are known to affect profoundly the odour of the compound in which they occur. These isosteric substances differ from the compounds they contaminate only in lacking one link of the chain, or in having one different link, or in both these respects. They do not differ in the number of free electrons in the nucleus. A good and very well known example of compounds of this kind is the pair benzene-thiophen. Benzene from coal tar, unless specially purified, contains small proportions of thiophen which have a pronounced unpleasant effect on the odour of the benzene. The structures of the two compounds are indicated in the following formulae:

BENZENE

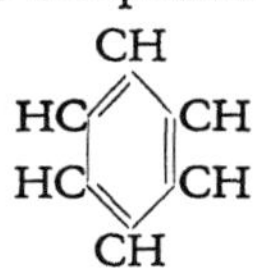

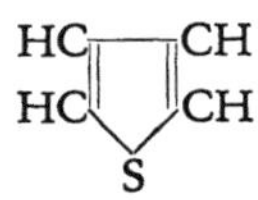

THIOPHEN

Guillot asked himself the question: if a minute proportion of an isosteric substance be added to the highly purified other member of the pair, will the change produced in the odour and the difficulty encountered in separating the two substances be such as to suggest some special relationship between the isosteric pairs not shown by comparable mixtures of non-isosteric substances? He then prepared the following mixtures and subjected them to fractional separation, plotting curves showing the progress of the process:

benzene contaminated with pyrrole
benzene contaminated with thiophen
benzene contaminated with pyridine
benzaldehyde contaminated with furfuraldehyde
naphthalene contaminated with indole.

The first four of these mixtures were fractionally distilled whilst the components of the last mixture were separated by fractional crystallization. He does not tell us much about the changes in odour observed during his experiments. He found, however, that the addition caused an increase in odour and he showed, quite clearly, that as regards separation, the mixtures do not differ from comparable non-isosteric mixtures. His conclusion was that the changes in odour noted by the perfumers are *not* due to isosteric impurities which, moreover, are no more difficult to separate than are similar proportions of non-isosteric substances.

It has long been known that groups of substances more or less closely related to each other chemically, homologous series of organic compounds for example, often have similar odours, that certain chemical groupings (osmophores) are frequently associated with the existence of odour and that connections between odour and properties such as unsaturation and "residual affinity" can sometimes be traced. Amongst those who first discussed these questions were Gertrud Woker (1906), Austerweil and Cochin (1910), Durrans (1919) and Delange (1922 and 1943). The publications of these investigators are still worthy of study as are those of Niccolini (1934, 1940) who divided the chemical elements into the olfacto-positive (those yielding odorous compounds) and the olfacto-negative (those not yielding odorous compounds). Those interested in the subject should also study the many papers still appearing that deal with the odours of particular groups of compounds, devoting special attention to the outstanding contributions of Beets. The groups include the following: acetophenone derivatives, arylalkanes and other hydrocarbons, bicyclofarnesol derivatives, hydro-aromatic epoxides, ionone derivatives, lactones and butyrolactones, macrocyclic-compounds (decanolides having musk odours), pyridine derivatives, thiophen derivatives, and vanillin derivatives. As Klouwen remarked not long ago (1960), these investigations suggest that the quality of the odour of a compound depends largely on the steric structure and is modified, especially in the smaller molecules, by the presence of osmophoric groups. Together with his co-worker Ruys, Klouwen has also given a striking illustration of this by showing that the very well known similarity in odour between benzaldehyde and nitrobenzene is to a great extent retained in their alkyl-substituted derivatives.

But, of course, it is also very well known that similar odours are found amongst substances having quite different constitutions. The musk odour, for example, which occurs in quite a number of plants and in more than twenty animal species (birds, mammals, molluscs, reptiles and at least one insect) is also encountered in synthetic chemicals from six or more chemical classes. In this connection, it is particularly interesting to note that when Naves (1951) prepared pairs of odorous substances from the same starting materials, one of each pair containing deuterium in place of hydrogen (*e. g.*, phenylethyl alcohol with OD in place of OH, methyl anthranilate with NHD and ND_2 in place of NH_2) he found that the deuterium compounds had odours distinctly different from those of the corresponding hydrogen compounds and he suggested that natural odorous materials from flowers may have different isotopic compositions and hence may smell differently. However that may be, it is at least not impossible, as Stoll has remarked that "there exist some molecules of special form and electronic system greatly favouring the production of odour. They constitute the real perfume substances, so actively sought by the perfumer".

Although Stoll's special forms would obviously not include all optical isomerides, the question of odour differences accompanying differences in the spatial arrangement of molecules is of peculiar interest from a chemical as well as from a biological point of view. About seventy years ago, it was claimed that some terpenes did exhibit such differences and German and Indian workers later affirmed that differences of intensity, at least, had been detected in some cases. However, other workers maintained that no such differences exist and it is quite certain that no case is known where one enantiomer is odorous and the other devoid of odour. As a result of the studies of Doll and Bournot in Germany, Guillot and his co-workers in France, Posvic in the U.S.A. and (especially) Naves in Switzerland, it now seems probable that differences of intensity, which are sometimes of considerable magnitude, do occur. Guillot regards this as beyond possible doubt. And it is at least possible that there are also qualitative differences. Naves insists that trustworthy results can be obtained only if the pairs of isomerides (*e. g.*, of menthol, dimethyl-*trans*-hexahydrophthalate, *iso*borneol, camphor derivatives, methylhexylcarbinol) are prepared from the racemic form and this form should later be reconstituted and its odour compared with that of the pre-resolution racemate. He maintains, also, that odour comparisons should be made at minimum perceptible level and he stresses the paramount importance of olfactive purity, a condition, it may be added, immensely difficult to attain. And he believes that progress in knowledge of the mechanism of olfaction and of olfactory physiology is most likely to come from stereochemical research.

However that may be, there is already a substantial amount of evidence suggesting that it is not so much the composition and constitution of a molecule as its magnitude and form that determine the quality, if not also the strength of its odour. "Even the senses of taste and odor", says Pauling, "are based upon molecular configuration rather than upon ordinary chemical properties — a molecule which has the same shape as a camphor molecule will smell like camphor even though it may be quite unrelated to camphor chemically . . . The specificity of the physiological activity of substances is determined by the *size* and *shape of molecules* rather than primarily by their chemical properties." This was the view of Timmermans also. He stated that the various diverse compounds having camphor-like odours have spherical molecules which have low entropy of melting — they are "globular" compounds and hence, although very different in chemical constitution, have many common properties, odour similarity being one.

In this connection, it is of peculiar interest to note that the inorganic compound borazole, which has a molecular weight and structure very similar to that of benzene also has a very similar smell. The analogy of electronic configuration parallels the analogy of odour.

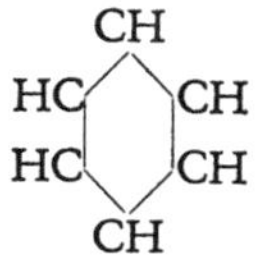

Benzene
Mol. wt. 78

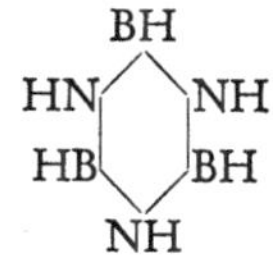

Borazine
Mol. wt. 80.5

Molecular shape or "profile" is also one of the bases of the very promising profile-functional group theory of Beets who calls attention, not only to the importance of the profile, but also to that of the various functional groups (osmophoric groups) that have repeatedly been shown, at least in homologous series, to have a definite influence on quality and intensity of odour. Possibly too, as Stoll has suggested, the ratio of the molecular weight of the osmophoric groups to that of the whole molecule may be an important factor, so that, if the ratio varies beyond certain limits, the odour may disappear. In this respect, the molecule is, perhaps, to be regarded as a piece of sculpture. If it should eventually be found that the olfactory receptors have a variety of shapes as proposed by those who support the "site-filling" theory of olfaction, this view of the significance of molecular shape would be further strengthened.

It is perhaps of interest to note here that, as long ago as 1933, Hukui plotted the melting and boiling points of 188 inorganic and 898 organic compounds, noting those that were odorous. He found that the values for all the odorous compounds fell within a triangle containing boiling points less than 600° absolute. The few non-odorous compounds represented in the triangle were non-surface-active and non-volatile. Fifty compounds that he examined for surface tension in aqueous solution could be arranged in five groups in which the odorous were separated from the non-odorous.

The Race and the Individual

Little is known about racial differences in olfactory acuity — if such differences in fact exist. — and, as we have seen, the observations of Rivers "lend no support whatever to the idea that the acuity of smell is greater in savage than in civilized races". If the statements of the Japanese writer Adachi are trustworthy, however, there are certainly differences of some kind. To Europeans, his remarks are not a little humiliating. Giving them the title "The Odour of Europeans" he goes on to claim that, although Europeans themselves appear to know nothing of their characteristic pungent and rancid odour, this is nevertheless well known to others. At the time he wrote (1903), the odour of Europeans had never been scientifically studied. Children and old people are almost free from it, he says, but it is particularly strong in women. Asiatics, he adds, seldom have such odour which originates chiefly in the armpits and, if it is found that a Japanese young man has "armpit" odour, he is (or was) exempt from military service. These and other comments in Adachi's paper need not be further examined here, especially as they consist entirely of unsupported statements by himself. That they have some basis can scarcely be doubted, however, and it is certainly the case that, at one time, medical men and others in France not infrequently referred to the "odour of the English" which some of them described as very persistent and long-lasting. This odour was attributed, in some cases at least, to the English diet (of mutton, for example).

It has been shown, too, that malaria mosquitoes detect individual odour differences between human beings, their perspiration probably being the controlling factor.

Whatever the truth about racial differences in olfactory acuity, it is very well known, as we shall again see, that very great individual differences exist just as there are certainly age and sex differences. One example of astonishing acuity may be given here. It was recorded some time ago in a medical journal. "Some may train one sense more than another", says the writer, and continues, "Professor von Neusser's acuity of smell was so pronounced that he frequently diagnosed a case on the odor of perspiration. He would walk into a ward of the general hospital in Vienna, lift a patient's arm, sniff and name the malady; then he would proceed with a thorough examination and discussion from every possible point of view. Rarely was his firs; impression proved wrong by the exhaustive study."

Nomenclature

Les odeurs ne sont pas faciles à définir, peut-être parce que leur nature n'est pas trop bien connue, je les appelerois volontiers "des sels volatils enveloppés de souffres très exaltés, seuls capables d'affecter par leur subtile activité le sens de l'odorat d'une manière qui leur est propre". *Poncelet*

All compounded tastes and smells are also made up of the simple ideas of these senses. But they being such as generally we have no names fᵒr, are less taken notice of, and cannot be set down in writing; and therefore must be left without enumeration to the thought and experiences of my reader. *Locke*

The study of olfaction has been very greatly hampered by lack of an appropriate nomenclature and a consequent absence of percision that is almost complete. Even to define odour, as Sagarin has shown, is a difficult matter and, as long ago as 1879, T. H. Huxley pointed out how popular views on the subject were (they still are) metaphysically unsound.

As we have seen, the list of odours drawn up by Teuber contained 400 terms. These could be divided into the three following groups: pleasant or unpleasant odours; odours similar to those of plants or animals; and odours similar to those of chemical compounds. It may be added, too, that difficulties of nomenclature are accentuated by language difficulties. The French description of a perfume, "sa tenue est moelleuse", for example, has been translated (jokingly of course) into English thus: "its dynamism is mellifluous!"

By any other Name as Sweet

O, how much more doth beauty beauteous seem
By that sweet ornament which truth doth give.
The rose looks fair, but fairer we it deem
For that sweet odour which doth in it live.

Shakespeare

Oh, wer um alle Rosen wüsste
die rings in stillen Gärten stehn —
oh, wer um alle wüsste, müsste
wie im Rausch durchs Leben gehn. *Morgenstern*

Crocker and Henderson, as we have seen, would have us express the odour of the rose by the number 6423 and rose-like perfumes by similar four-digit numbers. Aronsohn, on the other hand, suggested that, for rose-like we substitute the word *cihroic.* At an international conference, held in 1894, he explained how he came to construct this and other similar words intended to describe olfactory qualities. Like many others, he had felt the great handicap imposed by the lack of an appropriate terminology for odours. Unlike the others, however, he proceeded to invent a vocabulary — probably the only one that has ever been invented for this purpose. Having shown, he believed, that each of the various fibres of the olfactory nerve responds to the action of one type of odour only (the "specific energies" theory) he concluded that all odours for which an appropriate perceptive element existed amongst those fibres, must be of the same nature and hence could be regarded as constituting an olfactory class. Accordingly, he based his vocabulary on chemical formulae and, in his lecture, gave examples of its application. As a first step in transforming these formulae into words, he replaced the numbers used in the formulae by letters chosen thus:

1 = a, 2 = b, 3 = c, 4 = ch, 5 = d, 6 = e, 7 = f, 8 = g, 9 = h, 10 = i, 11 = k and so on. In this way the formula for camphor, for example, is changed from $C_{10}H_{16}O$ into CiHpO or cihpo and the term for camphoraceous becomes cihpolic. No rule is made as to the order in which the various atoms in the compounds are to be arranged but, in the examples given, the order for organic compounds is C, H, O. Where the transformed formula consists of

consonants only, or where successions of consonants would render word-formation difficult, appropriate vowels are introduced, these vowels being changed to indicate "shades" or "nuances" within the olfactory class and to express qualities such as pungency, fragrance, putrescence, and so on. As endings for the words obtained, the usual adjectival endings employed in the language in use are chosen, thus English -ic, French -ique, German -ig or -lich. In this way, the odour of the rose, due chiefly to geraniol $C_{10}H_{18}O$, is said to be cihroic, that of phenol (carbolic acid) C_6H_5OH is cedohlic, that of hydrogen sulphide SH_2 is sahbic, and so forth.

It is at once clear today, if it were not already so in 1894, that this terminological essay suffers from many "inexactitudes". It takes no account, for example, of the well known fact that substances having the same empirical formula may have entirely different odours whilst others having similar odours may have quite different formulae. The inventor himself may have used his vocabulary. It is unlikely that anyone else ever did. At the best, it can be regarded only as a curiosity.

Science?

Lower Sense

> Perfumery has never given rise to an art of any pretension. Moreover, artificial essences never yield the same pure and delicious fragrance as natural flowers and fruits. There is always a sickly tinge about their sweetness. This inability to compete with nature is a fatal objection to perfumery as the basis of a fine art. It is characteristic of the true aesthetic arts generally that they are *more* beautiful than nature, because they gather together all that is lovely, and omit all that is low, discordant or ugly. Thus the Discobolus or the Medici Venus is more beautiful than any living nude human figure, because it combines all the best points of many; a landscape painting is lovlier than reality because it excludes all unpleasant accompaniments; a great poem takes us into a region of ideal delights; a grand oratorio immeasurably surpasses any natural collection of sounds. But a lily-of-the-valley or a heliotrope is worth all the mille-fleurs or frangi-panni ever manufactured.
>
> *Grant Allen*

> Il est banal de dire que l'élaboration d'un parfum nouveau est assimilable à la création de toute oeuvre d'art. En realité il s'agit de quelque chose de très particulier, et l'élaboration d'une formule de composition est certainement beaucoup plus empirique que celle de n'importe quel autre type d'oeuvre artistique ou artisanale. Le parfumeur est entouré d'un nombre considérable de substances odorantes, utilisées par lui comme matières premières; il les mélange, en suivant sa seule inspiration ou sa seule expérience, et le mélange, auquel il aboutit est soumis seulement à sa propre appréciation olfactive.
>
> *Guillot*

A few words should be said here about Von Skramlik's *Handbuch der Physiologie der niederen Sinne*, a work which is surely the best in its field but seems to be greatly neglected perhaps because it appeared nearly forty years ago.

Perhaps the most interesting parts of the book are those dealing with odorous mixtures and with the number of odours known. (Meunier states that a natural perfume may have as many as 300 constituents.) Of all the known (in 1926) chemical compounds, organic and inorganic, perhaps one-fifth or thereabouts (that is to say, rather more than 60,000) must, he held, be regarded as odorous. Nevertheless, he also said, there have so far been discovered only about fifty substances having odours which are free from non-olfactory qualities. These fifty substances are said to be *purely odorous.* (Purely odorous substances are those which cannot be localized in one nostril rather than in the other when held under one nostril while an odourless substance is held under the other.) And there are also only about fifty purely odorous qualities known. Thus there are just as many odour qualities as there are purely odorous substances. (It has been pointed out by several authors that the number of odour qualities, like the number of tastes, may actually be much smaller. If so, then von Skramlik's "pure odours" are, in fact composite.)

His discussion of the odours of mixtures is unique and of the highest interest. In such mixtures, he found, there are "at both ends" zones where one odour so preponderates that the other (or others) are completely suppressed, predominance depending on concentration and varying with changes in it. Between the terminal "zones of suppression" there are "zones of succession" where odours compete for attention and are perceived successively. Here also are "zones of concurrence" where two or more odours can be distinguished together simultaneously. It is never possible, he maintained, to produce a "unity" (*i.e.*, a non-heterogeneous odour) by mixing two or more pure odours although very close imitations of a pure odour can be made. When five or more odorous substances are mixed, the olfactory identification of the individual constituents becomes extremely difficult and, at most, only two or three constituents can thus be detected.

Art?

La composition des parfums exige certes un minimum de dispositions naturelles, de l'intuition, de l'imagination, l'éducation constante du goût et la foi dans un idéal, mais elle est, avant tout, *un metier,* un métier exclusif qui nécessite du temps, de l'application, de la persevérance. On s'y perfectionne laborieusement et là, autant qu'ailleurs, l'expérience est une somme d'inexpériences. C'est un métier, au sens noble que l'on s'efforce de rendre à ce terme qui ne doit pas effaroucher les vrais "compositeurs de parfums", ceux qui connaissent les longues méditations sur les milles problèmes posés par l'infini des combinaisons, comme sur l'enchevêtrement des incidences à la prévision desquelles on aspire; ceux qui patiemment, jour après jour, se penchent sur les odeurs pour étudier leurs affinités, pour tenter de saisir quelques-uns des secrets de leurs comportements et qui savent quelle accumulation d'observations représente parfois un modeste gain; ceux qui tormente jusqu'à l'insomnie la recherche d'un accord ou la poursuite d'un souvenir olfactif. Ceux-là savent bien que composer des parfums est un métier qui ne souffre pas le dilettantisme et que seule la méconnaissance de cette vérité a permis à tant d'erreurs, voire à tant de sottises, de s'accréditer.

Roudnitska

If the enquiry is made as to why there are no recognized arts of savours and scents corresponding in rank to the arts of forms, colours and sounds, the answer, Colvin tells us, is that taste and smell are unintellectual and therefore lower senses, or, alternatively, that these senses yield only private pleasures which cannot be built up into separate and durable schemes of which everyone may have the benefit since they cannot be monopolized or used up. And Grant Allen said long ago that "no fine art can be based upon odours, for the human race at least. There are no associated emotions upon which the art could play. Colvin suggests, however, that there may nevertheless be an "art of scents" since "in Japan, it seems, there is a recognized and finely regulated social art of the combination and succession of perfumes".

This social art has been described by Henning who believed that modern Japanese know nothing of it. It is a ceremonial scent "game" known as *kôdô* which, he said, originated before 1151 A. D. and was "played" in tea houses, often under the direction of masters of the tea ceremony. There exist, apparently, only a few descriptions of the procedure in the Japanese language. Henning succeeded in obtaining the apparatus and materials used in the ceremony from the German Consul-General in Nagasaki.

The ceremony proceeds as follows. The master of ceremonies, in an exactly prescribed form, drops certain scents which the participants have to determine, into an incense-burner. The first participant, having received the jar containing the smouldering perfume, sniffs in order to facilitate recognition of the odour and then ceremoniously passes the jar to his neighbour. In this way, each participant in turn is allowed to test his powers. Usually ten specimens representing four varieties of incense are handed round thus. To each participant there are given as many decorated tablets or counters as there are specimens. These counters, which are gold lacquered, bear the design of a leaf, a different leaf for each participant. When a participant has determined a scent, he writes the name on a slip, encloses this with one of his counters in an envelope and passes it to the master of ceremonies. At the end of the ceremony the master announces which "leaf" has won (it would not be polite to give the name of the winner) and the prize is awarded. All the scents used are of the incense type. A vessel containing glowing charcoal is employed to ignite them. Almost all of them were imported from India since Japan is poor in aromatic woods. The delicate apparatus of lacquer ware which is used includes tweezers, wooden rods, spoons for ashes, feathers for brushing and so on. Henning points out that, in such a ceremony, it must be far from easy to determine odours especially as no help is derived from the sense of sight. Moreover, there is often great similarity among the odours and, after a time, the whole room becomes filled with the mixed aromas. The Japanese government, it seems, encouraged participation in kôdô as a means of diverting the people, gathered in tea houses, from 'dangerous' political discussions.

Henning seems to have been unaware of the fact that several other European accounts of kôdô had previously been published. One of these was written, about twentyfive years earlier, by Dillon who examined a complete set of the equipment at the Burlington Fine Arts Club. Dillon's description differs somewhat from that of Henning. It is less detailled but no less interesting. According to Dillon, the first mention of the incense ceremony appears in a Xth century romance.

The distinguished poet and writer, Sir Edwin Arnold (1832—1904), who suggests that the ceremony might very well be introduced amongst people of education and fine sensibilities in Europe, points out that the Japanese had fewer perfumes than had the Europeans but that they appreciated intensely those which they had and gave them lovely and appropriate names, for example, "breath of twilight" and "dew on the mountain path". Sometimes, playful or fanciful names were given. Only solid perfumes, often in powder form, were used. The number of participants in the ceremony varied considerably but seems often to have been between three and six. Frequently (or always) they were nobles or aristocrats and, in any case, they were persons of cultivated tastes. For a day or more before partaking in the ceremony, they refrained from smoking, drinking tea or saki, eating scented sweetmeats, or applying oil or perfume to the hair. They wore new, or at least unsoiled clothes and, before entering the room in which the ceremony took place, they laid aside outer garments which might have acquired odours in the street or elsewhere. The master of ceremonies and his assistants (Arnold says that the master of ceremonies always had a lady assistant but others do not confirm this) allowed the participants to smell the fumes of known specimens of the varieties of incense used before the actual session of determination began so as to provide olfactory memories. But one incense, known as the guest incense, remained as an unknown. The counters or tablets indicating the guesses were either dropped, "in silent ballot", into a casket or placed on the appropriate square of a chequer board provided for the purpose.

Complementary to Arnold's account is that of the great authority on Japanese life, Lafcadio Hearn (1850—1904). He tells of varieties of the incense ceremony much more elaborate than the tea ceremonies, of the exquisite beauty and finish of the equipment, especially the incense-burners which were often great works of art, and of the writing utensils and books in which records of the results of the ceremonies were preserved. Lacquer, of course, was not the only material used for the equipment: there were also magnificent pieces made of carved wood, porcelain and metals — iron, bronze, silver, gold. In addition, he gives a translation of one of the records showing that in a ceremony in which ten tests were made with three known varieties of incense and a guest incense, the winner, Young Pine, made only two mistakes whilst White Lily was correct only once.

Henning's belief that the modern Japanese knew nothing about kôdô was certainly not entirely correct since Buntarô Sugimoto published a 400-page book on the subject in 1936 and Senkichirô Katsumata devoted one chapter of his book "Gleams from Japan", issued in 1937, to "Incense Ceremonies".

Katsumata tells us that kôdô had been a favourite pastime of the aristocracy in ancient times but had gone out of practice. It had been the custom, he says, for each participant in the ceremony to contribute various self-prepared incense compositions to the ceremony. However, in 1919, it was again finding its votaries among the ladies and gentlemen of leisure and taste. At one time, he adds, celebrated masters of the ceremonies established schools where they gave instruction in the art and he refers to one who became so expert during his prime that he never failed to identify the aromas presented to him. But in old age, his skill declined. (Incidentally, Katsumata also relates that Kyoichi, a celebrated maker of *bonkei* or miniature landscapes constructed one including an execution ground. In order to attract customers to his shop when these bonkei were displayed, he burned a material giving an odour similar to that of burning corpses since the corpses of beheaded criminals were cremated. Thus was added "an olfactory feature to intensify the effect".)

That Europeans, whose pastimes included such games as roulette, faro and poker, should ridicule kôdô may seem surprising today but they certainly once did so, calling it, for example, "an innocent, not to say insipid, *jeu de société* such as might suggest itself to a party of schoolgirls" and suggesting that when it was in vogue, Japan was "in her childhood — her second childhood". It is to be presumed that those who held such views were also unaware that the Japanese had invented an ingenious incense-"clock" *(Koban-Dokei)* which made it possible to 'smell' the time. In this instrument, small tablets of incense, marked with Chinese zodiacal signs, burned down in succession, one by one. Since each tablet emitted a different aroma, people with sensitive noses could thus tell the approximate time. In old temples, this method of time measurement was still practised not very many years ago, particalarly in connection with the reading of *Sutras* and perhaps it is still practised. Bedini has recently given an account of various types of incense or fire clocks in an article entitled "*Holy Smoke*".

No serious student of sociology would now make the mistake of supposing that the incense ceremony was merely a rather elaborate game having for its purposes amusement and the development of skill in the identification of odours. On the contrary, it was closely connected with many kinds of symbolism. It was also, as Brinkley has shown, "a literary pursuit, designed in great part for testing the players' knowledge of classical poetry and their ability to apply their knowledge. For a literary name had to be given to each incense identified and great erudition was often exercised in choosing the names".

Together with the tea ceremony and the art of flower arrangement, kôdô formed an essential part of the finished education of the aristocratic ladies and gentlemen who alone practised it. Lieut.-Colonel Strange, whose illustrated account is perhaps the fullest description in English which we possess, tells us indeed that kôdô "was essentially a highly elaborated form of intellectual culture, framed and garnished with a charming ritual and, most curious and significant of all, based on the cultivation of one of the senses — that of smell — which has absolutely no place whatever in Western aesthetics. Yet it must be within the experience of every one of us how strangely that neglected and dormant faculty will sometimes awake and bring back memories of long-forgotten scenes and of visions that seemed to have passed forever beyond the ivory gates."

(An illustrated account of "the ancient art of incense in Japan" as practised today was published in "Soap, Perfumery and Cosmetics" in 1965 by the president of the Japan Perfumery and Flavouring Association, Dr. Teikichi Hiraizumi. Curiously enough, this account does not include the word kôdô nor does it refer to any other Japanese or European description of the ceremony.)

Measurement

Because of the severe limitations of organoleptic procedures, more or less successful attempts have often been made to replace them by processes affected only slightly or not at all (if possible) by subjectivity. Minute quantities of odorous substances (and hence, intensity of odour) can be determined by various forms of chromatography; by measuring the change produced by such substances in the diameter of a pendant drop of water, oil, or mercury; by determining the change in electric potential that occurs between two surfaces (*e. g.*, those of a gold disc and a vibrating membrane coated with a gold film) when airborne vapours are differentially adsorbed on them; by measuring electrically the change in temperature that occurs when odorants pass over two thermistors one of which is coated with adsorbent material such as casein, cephalin, polyvinyl chloride and peanut protein; by determining the amount of adsorbent (*e. g.*, activated charcoal) required just to suppress the odour in a current of air passing through it or by employing an "osmpile". This device, claimed to be "more sensitive than the human nose", is a "simple receptor capable of replacing the human nose in the detection and, especially, the determination of *certain* odorous substances". (Such substances are: alcohols, aldehydes and terpenes to which osmopiles are extremely sensitive. Osmopiles are quite insensitive to lactones and musks.)

Essentially, an osmopile is a device in which the vapour of a substance is adsorbed on an aqueous surface (water or salt solution) located between two electrodes, the resulting change in the electric current passing between the electrodes being measured. Osmopiles, however, do not respond only to substances that are odorous and it is not certain that they are always more sensitive than is the human olfactory organ.

Some degree of objectivity in measuring can also be attained by determining the effect of odours on the dilation of the pupil; on the flow of blood (in the finger tip); on the respiration (Hongô claimed, in 1936, that the acceleration of respiration produced by odorants parallels in extent and duration the concentration of the materials when they are sniffed by rabbits); on the salivary reflex or on the electro-encephalogram.

Another olfactometric method which might be made objecive is the psychogalvanic procedure. Although it has yielded only doubtful results, it is perhaps worth mentioning here.

Since nervous emotion is accompanied by measurable changes in the electrical conductance of the skin, methods of measuring such emotion have been developed. These methods are very sensitive and the effects which they measure are largely sub-conscious. The apparatus used consists, essentially, of suitable resistances connected with cells and a galvanometer by means of which the changes of conductance are registered. Contact is made with the palm of the hand or the forearm.

In 1923, Auld, who wished to obtain "an actual measure of smell with a view ultimately to getting a permanent and possibly quantitative record of the odours of various substances", spent some time endeavouring to apply the "emotion meter" to his purpose. He found, for example, that very striking effects were registered when letters, agreeable and disagreeable, were received. The effects of very slight pain, such as that caused by a tiny pin-prick, and even those of the anticipation of pain which did not eventuate, were very marked. Similarly, very pronounced effects were registered by his three subjects (one sub-sensitive, one normal and one super-sensitive) when discriminating between pleasant and unpleasant odours. None the less, the object of the tests was not achieved. The slight differences in odour of petroleum products were found difficult to record and, when the apparatus was rendered more sensitive in an attempt to detect slight olfactory preferences, outside influences became so marked that the procedure was made useless. Believing that the whole process was then too much in embryo, Auld abandoned his attempt but he pointed out that satisfactory results might possibly be obtained if the subjects were emotionally insulated and rendered bodily inert and if it should prove possible to eliminate outside influences altogether.

The short report on the psycho-galvanic reflex produced by olfactory excitation published a few years later by Bijtel and Van Iterson, did not entirely confirm Auld's views. This report indicates that curves expressing the changes can be plotted. The authors used camphor, citral, caproic acid, glacial acetic acid, ammonia, cinnamic aldehyde, and cigar smoke. They found that quantitative differences could readily be recorded but that there was considerable uncertainty about qualitative differences. Individual differences were frequently encountered and very energetic responses were elicited by some excitations (these produced by glacial acetic acid, for example, were so strong as to cause pain).

These British and Continental researches were followed by an American psychological study in which one hundred adolescent children were examined, fourteen odorous substances being used. The investigators, Coombs and Shock, divided the odours into three groups — pleasant, least unpleasant and most unpleasant — and also studied odorous strength. The tentative conclusions which they reached were: first, that the affective quality of the stimulus must be considered as well as the affective strength as a factor determining the magnitude of the galvanometric deflection and, second, that for equal degrees of affective strength, the unpleasant stimuli tend to give greater deflections than do pleasant stimuli.

A report on psycho-galvanic procedures has also appeared in France where doubt was rightly expressed concerning the scientific status of the work concerned. The subjects were film stars from Hollywood and the instruments employed included the "psychogalvanoscope" and the "electro-oleo-convector". The experimenters were Waugh and Read. Using Hathaway's psycho-galvanoscope and Read's olfactometer they sought to determine which perfumes served best as "evocators of past emotions". The significance of these studies is clearly rather doubtful. Perhaps their chief value lay in the delightfully ironic comment elicited from the French writer who described them in the journal *Parfumerie Moderne*. He apparently felt obliged politely to hint that the secret desire of the Americans to outdo the European perfumers could not succeed. "Nous serions fâchés de jeter une ombre", he wrote, "même legère, sur ces illusions juvéniles; l'Europe n'est pas mûre pour les révélations de l'électro-psycho-olfacto-galvanoscope, pas plus que celles de l'électro-oleo-convector. Les Françaises, notamment, n'ont besoin d'aucun outillage à lampes d'amplification pour vibrer plus ou moins agréablement sous l'influence de tel ou tel parfum qu'aucun appareil ne leur fait choisir: leur goût affiné leur suffit."

Much later, olfactory psycho-galvanic tests were conducted in Italy by Semeria, in the U.S.A. by Borsanyi and Blanchard and in England by Moncrieff but, in all cases, the results were disappointing as were also those of Galli and Incutti (1964), Asaka (1965), Rous and Synek (1966) and Neuhaus and Goldenberg (1966). The latter examined changes in the human galvanic skin response by olfactory stimulation with butyric acid, skatole and muskinone and found that reduction of resistance occurred.

An additional process that may be of value in olfactometry is the electro-physiological method (first applied to olfaction it seems, by Hasama, in 1934) which has been successfully used in olfactory studies in recent years especially by Adrian and his colleagues. Olfactory stimulation has been shown to produce an effect on the electro-encephalogram and, in experiments on rabbits and other animals, it has been found that, since the olfactory receptors may vary in relative sensitivity to different smells, different olfactory stimuli may produce different distributions of excitation so that "a familar smell, like a familiar sight, is recognized by the specific pattern which it arouses in the brain" (Adrian). However, direct electro-physiological investigations of olfaction in human beings can usually not be permitted (psychotic patients only seems to have been examined thus) and, so far, the factor that has been chiefly studied has been *quality* not *intensity*. Moreover, it appears that non-olfactory stimulation (*e.g.*, acoustic and visual) may sometimes produce olfactory-like electro-physiological responses and that, even in the field of quality, the results obtained may "cut a swath across most odor classification schemes" (Gesteland). Le Magnen has pointed out, too, that the electro-physiological method is not yet sufficiently advanced to permit analysis of the olfactory discriminatory process and Mozell has confirmed this although he also claims that the magnitude and duration of the electrical response can, to some extent, be correlated with the concentration of the odorant involved.

All these methods, however, are lacking in precision, sensitivity or specificity. Castello's "electronic nose", for example is not a true olfactometer but an analytical instrument in which gaseous substances, especially halogenated compounds, are ionized and determined, the unit of measurement being the "osm". Organoleptic methods, in fact, are almost certainly indispensible and are usually, if not always, much more sensitive than are other procedures. It is with them and not with instruments or methods that measure both odorous and odourless substances that final decisions are to be made.

If then odour intensity is to be measured, however subjectively, it must be measured with the nose (usually the human nose) so that, at the same time, the sensitivity of that organ is also determined, the magnitude generally concerned being the "minimum perceptible". Thus, for example, von Skramlik has made a list of purly odorous substances (*i.e.*, substances that affect the olfactory sense only) and of "ordinary" odorous substances *(i. e.*, substances that affect other senses in addition to the sense of smell), giving, in each case, the minimum perceptible (in millionths of a gramme per 50 c. c. of inspired air) and the corresponding number of molecules. This number which is 2.1×10^{17} for *iso*amyl acetate and 5.4×10^{8} for trinitro-*tert*-butyltoluene (one of the synthetic musks) decreases within certain limits as molecular weight increases.

It is obvious from these figures that, however minute may be the minimum *weight* of odorous substance required to produce a a sensation of smell, the corresponding *number* of molecules is nevertheless enormous (at least hundreds of millions). With the nose, impurities added to an odorant can be detected even in proportions not detectable spectroscopically and Sacher has shown that it is possible to determine by smell, with fair accuracy, the end point in the volumetric estimation of, for example, *iso*valeric acid using sodium hydroxide.

Although an instrument (olfactometer) is frequently used in determining the minimum perceptible, this magnitude can also be determined without any instrument. Consider the procedure adopted in 1886, for example, by Emil Fischer and Franz Penzoldt. Of the many determinations of the olfactory minimum perceptible that have been made none is more interesting than this. These workers were not only amongst the earliest in the field. They also chose a method (it was also employed, many years later, by Schley and by Baldus) in which the experimental conditions resemble those most likely to be encountered in everday life more closely than do the conditions prevailing in other procedures. Having conducted some preliminary experiments in order to assist their selection of odorous substances, they chose mercaptan[1] (presumably ethyl mercaptan, C_2H_5SH) and one of the chlorophenols ClC_6H_4OH. Of each of these, a stock solution was prepared by dissolving one gramme in a litre of pure alcohol, the solution being then diluted as required. The weighing of the substances and the diluting and the cleaning of the outside of the vessels to be used in the room were carried out by a person not concerned in the investigation, in a laboratory far removed from that room. It was an empty room of 230 cubic metres content with plaster walls and a stone floor. After Fischer had entered the room, it was totally closed and he vaporized a known quantity of the odorous substance, using a small hand pump. Then the air in the room was thoroughly mixed with a large fan. Finally, after a given time Penzoldt entered and tested the odour by sniffing. As a check, a test was also made by an independent observer. When estimates had been made of the volume of air inhaled at a single sniff (250 c. c.) and of the volume passing through the olfactory cleft (50 c. c.), it was possible to calculate the minimum perceptible. The results

[1] Blinova (1965) has recently determined the minimum perceptible of ethyl and other mercaptans.

were astonishing. As little as 1/4,600,000 mg of the chlorophenol could be detected and as little as one-hundredth of this amount, 1/460,000,000 mg of mercaptan. Commenting on these extraordinary figures, Fischer and Penzoldt pointed out that, with their noses, they could detect mercaptan at a concentration which was approximately 250 times lower than that at which Bunsen and Kirchhoff, *using the spectroscope,* could detect sodium.

The "olfactoria" (capacity 500 cubic feet, for example) that have recently been constructed in the U.S.A. (and possibly elsewhere) are essentially the same as was Fischer and Penzoldt's test room. They are probably more reliable since they are made of easily cleaned material (glass or metal, for example) of low adsorbing power and are fitted with devices for introducing precise amounts of odorant, for circulating pure air, for removing odorous air, and so on. An olfactorium, however, is very expensive and much time is consumed in cleaning it. Zwaardemaker's glass-walled "camera inodorata" [2] to contain the head only is naturally much cheaper and perhaps no less reliable. When he used it, thresolds were lowered by as much as 34%. A similar apparatus was used at a much later date by Bernice Wenzel.

Fischer and Penzoldt's results show, once again, how extremely small are the the amounts that can be detected by smell. It is said, for example, that one gramme of iodoform, a compound having a very powerful odour, loses weight at the rate of about one milligramme in 100 years. Whether or not this is so, Aitken and, later, Bazzoni, who worked with natural musk found no support for the view, expressed by some proponents of radiation theories, that the material emits its aroma without losing weight. Aitken drew filtered air over musk into a flask containing a little water. When the pressure of air in the flask was then reduced by applying suction, no cloudy condensation occurred although such condensation was always observed when solid particles *(e. g.,* dust) were present. In confirmation of this, Aitken also showed that the odorous material emitted by the musk was not trapped by passage through a plug of cotton-wool. It follows from these results that if odorous matter is indeed emitted by solid musk, this matter is in the form, not of solid particles, but of vapour or gas. Bazzoni exposed a dried specimen of musk to a current of dry air during a period of seven months and, using a specially designed micro-balance, readily showed that a 14% loss of weight took place. (Natural musk, of course, is a complex mixture and hence the loss is due to evaporation of volatile matter all of which is not necessarily odorous.)

Those who make olfactory determinations, of course, must be in good health, free from nasal defects and aware that the threshold of *perception* is not identical with and is usually lower than the threshold of *recognition.* They should allow odorants time (not less than about one second) to produce

---

[2] This was the apparatus used by Ohma (see p. 115).

their effect and they must not allow themselves to develop olfactory fatigue. An experienced perfume chemist, it has been stated, can recognize by smell alone about 8000 substances and, according to Polak and Schwarz, he will distinguish four abilities in evaluating odours, namely, discriminating power, associating power, estimating power (quantitative) and reproducing power (*i. e.*, ability to repeat evaluation). Many people, especially if untrained, cannot readily identify odours. To them, the "triangle" test, *i. e.*, the test in which the odorant under examination has to be compared with two other odorants only one of which is identical with it, may with advantage be applied. A few people, on the other hand, seem to possess an extremely keen sense of smell. Such was the peasant described by Ebers in 1859. This "thief catcher" discovered stolen property, smelled out thieves, tracked human footprints, distinguished men from women and selected individual animals from flocks, all by his olfactory sense. He was examined by members of the Silesian Medical College who accepted his ability. Jean Carles, one of the world's most famous perfumers, on the other hand, has recently affirmed that "il n'existe pas, en parfumerie, de 'nez' privilégié". Everybody, he says, can have his sense of smell very much developed — it is simply a question of training. Whether or not this is true, it is certain that medical men, for example, can become very efficient in diagnosing disease by smell, being able, in some cases, to detect thus the approach of death even when pulse, temperature and feeling of well-being are not unfavourable. It is also true that olfactory acuity is sometimes reduced by disease (*e. g.*, diabetes, epilepsy and tuberculosis) and that "osmotherapy" may be valuable at least in some conditions (not in asthma). Perhaps it should be added that partial anosmia (*e. g.*, to acetone and to hydrocyanic acid) seems to be hereditarily transmitted sometimes, perhaps by women but not by men. As far as hydrocyanic acid is concerned, there is no doubt that up to 20% or more of human beings, both black and white, cannot smell it [3]. About one person in a thousand, cannot smell the extremely repulsive butyl mercaptan (skunk odour).

One of the first olfactometers, perhaps the very first, was that of Zwaardemaker who published his description of it in 1888 when it cost 1.5 Marks. He believed that it combined simplicity with accuracy. It consists of two tubes that slide one within the other, the inner tube being bent upwards at the outer end so as to fit into the nostril. The inner surface of the outer tube is coated with the odorous material. When the inner tube, which is graduated, is slipped into the outer one so as to cover its inner surface completely and air is drawn into the nostril through the inner tube, no odour is perceived since the air current is prevented from coming into contact with the odorant. By changing the position of the inner tube in relation to the outer one, more or less of the odorant is exposed to the

[3] In 1967, Brown and Robinette showed that there is no simple pattern of inheritance in ability to smell solutions of cyanide.

current and the intensity of the odour will vary. As a unit of measurement, Zwaardemaker chose the "olfactie". The intensity of an odour, he stated, is the reciprocal of the minimum perceptible which is determined by the partial pressure of the odorant in the odorous air and the olfactie is the partial pressure that corresponds to the minimum perceptible to a normal olfactory organ at average room temperature.

Zwaardemaker's instrument was later modified by himself and by others and he eventually produced a fairly elaborate double model. In many hands, the Zwaardemaker apparatus has probably provided results of considerable value (in 1927 Eykman used it to study large-ring cyclic ketones — musks — and odorants from Zwaardemaker's nine classes and earlier, in 1909, Hermanides used it and also the camera inodorata in determining the minima perceptible of eight of the nine standard odours whilst, in 1917, Huijer employed it in the olfactometry of aniline and five of its homologues) but it has at least one serious defect for the area of the exposed odorant surface is not necessarily proportional to the concentration of odorant in the inhaled air. Thus, if the air passing over the first part of the surface becomes saturated, no increase in the intensity of odour will results as the flow continues over the rest of the surface. Again, as Gavaudan and his co-workers have pointed out, Zwaardemaker's unit, the olfactie, "n'a qu'une valeur purement relative et empirique et parait n'avoir ni signification physiologique ni sens physique".

French investigators who followed Zwaardemaker dispensed with instruments. They allowed their odorants to evaporate in an enclosed space or dissolved them in a suitable (odourless or nearly odourless) solvent and tested them, after preparing a series of progressively diluted solutions, by sniffing beginning either with the most concentrated or, preferably, with the most dilute. (Many years later, it was shown that this simple "sniffing at bottle" method gives quite reliable results.) The odorants were either pure chemical compounds or natural mixtures such as essential oils.

In nearly all the many olfactometers, some of them quite elaborate, that have since been devised (Bernice Wenzel discussed many of them in her review of one hundred years of olfactometry) one or other of these methods of dilution with air (perhaps nitrogen, used by some investigators, is better) or with liquid, has been used. It has been claimed, however, that it is better to sniff at absorbent paper impregnated with the liquid than at the liquid itself.

Whatever olfactometric method is used, a considerable number of factors that may appreciably influence the result must be taken into account. If the odorant is present in a solvent or vehicle (oil, water, *etc.*) the effect of these must be considered, since solvents and vehicles often have pronounced although unpredictable effects on the odour. "Odourless" gases such as oxygen and nitrogen actually appear to possess odours when present at high pressure and this must be kept in mind. Account must be taken of the relative humidity and of the temperature at which determinations are made. Although, in nature, greater quantities of odorous material are released when humidity is high (*e. g.*, after rain) than when it is low, there is some evidence that odour intensity increases as humidity decreases. It seems possible that intensity increases somewhat with increase of temperature but since the temperature inside the nasal cavity almost certainly never varies outside a narrow range, changes in temperature are very unlikely to have any appreciable effect. Persons making olfactometric measurements should not have had medicaments administered to them recently because some drugs and other substances cause hypo- or hyperosmia [4]. They should probably not be smokers although some observers, experienced perfumers included, maintain that smoking has little or no detrimental effect on olfactory acuity and may even have a beneficial influence on it; and they should probably not recently have had a satisfying meal since acuity seems to be greatest before and least after such a meal. Fasting is said to increase acuity and obese persons (and criminals!) have been found to have low acuity.

When olfactometry is being practised, great care must be taken to avoid fatigue (adaptation) of the perceiving organ. Although olfactory fatigue may occur more slowly and be less pronounced than is usually supposed, it certainly develops fairly readily and its effect has been used as a means of classifying odours since like odours are readily or very readily affected by it and unlike odours are scarcely, if at all, influenced. It has sometimes been affirmed, also, that complete mutual cancellation ("compensation") of odours may sometimes occur. However, there seems to be no convincing evidence that such cancellation ever takes place except, perhaps, with some odorants when present in very low concentrations. (In 1921/23, Hofmann made some unusually interesting observations on fatigue, compensation and other related phenomena during a period of loss and recovery of his sense of smell.)

Odorants should be inhaled, preferably without inserting nose pieces. These have a disturbing effect. Blast injection is unphysiological and should not be practised especially as it has been shown that threshold concentrations are appreciably affected by the rate of flow of the odorous air current. With blast injection, the thresholds determined are not comparable to thresholds found in terms of concentrations of odorant and cannot be translated into molecular terms.

It should not be necessary to emphasise as strongly as possible that all odorants used in olfactometry should be as highly purified as possible and should have no trigeminal components. Purity in odorants is of importance, not only because of the disturbing effects of impurities but also because it has been found that, in mixtures, both antagonism and synergism may occur

---

[4] Administration of hyaluronidase, for example, increases human olfactory acuity and may even render "odourless" substances odorous.

so that the magnitude of the odour sensation perceived may be neither the sum nor the average of the individual intensities. It has also been found, indeed, that even when *all the constituents* (two, three or four) of a mixture are at sub-threshold concentrations, an odour can be perceived. Moncrieff claims that certain anti-odorants counteract the odours of "mixed" odorants (*e. g.*, acetic anhydride, formaldehyde) without counteracting their irritant properties. And, recently, Von Békésy has maintained that the direction of an olfactory source can be determined by using the time difference and the difference in sensation magnitude between the two nostrils as the air surrounding the odorous object is inhaled. But it is difficult to believe this.

Although it is probably impossible to prevent odorants from being adsorbed on apparatus used in olfactometry, steps should be taken to minimize or counteract the effect thus produced. About fifty years ago, Zwaardemaker and others studied the adsorption of muscone and other odorants on metal, glass, porcelain and other materials and found very great variations in the periods during which the odour could still be detected on them. Somewhat later, Hermanides, using the "nine standard odorants", found that these periods ranged from a few seconds or less to twenty days (skatole on steel) and Huijer studied the adsorption of aniline and five of its homologues on amber, glass and quartz. These findings indicate that the adsorptive effects of materials such as aluminium, stainless steel and, in particular, glass, used in the construction of olfactometers, must not be overlooked. Glass has been said, in this connection, to be the only material "free from inherent odour and readily cleaned" and Stuiver has proposed that errors due to adsorption should be eliminated by allowing the odorant to come into equilibrium with the walls of the apparatus before measurements are made.

It should also be noted that charcoal and silica gel, which are often used to deodorize air for use as a diluent, may be unsatisfactory because, it is said, they have odours of their own. Perhaps it may be added that rubber and other organic materials should be avoided in olfactometry because they are liable to decompose or to react with the odorants. And, obviously, environmental odours must be excluded as for as possible. (Amirov has found that, in 10% of healthy persons, olfactory sensitivity is considerably diminished by previous exposure to subliminal olfactory stimulation — with camphor, for example.) It should be remembered, too, that results may differ according to whether one nostril is used or both (separately or together). In fact, the dirhinic threshold, as theory suggests, is lower than the monorhinic.

It is obvious that, in olfactometry, a high degree of accuracy cannot be attained and it has been suggested that, isotopic tracers, especially labelled sulphur hexafluoride, might be used with advantage to increase accuracy (and, no doubt, to solve other olfactory problems) [5]. Barail has claimed that, with his instrument, an osmometer or osmagraph "accurate measurements of the intensity of all odors, faint or strong" can be made and that his is "the first instrument wherein the elimination of variables has been so perfected that the testing of an odor can be made by several trained technicians with variations of less than 5%". He also claims that 300 stages of intensity can be measured, although, in practice, about fifty are enough. However, it is doubtful if more than a very small number, possibly four or five, of intensities can be identified even by practised persons, and when both intensity and quality are varied simultaneously, the accuracy of intensity identification is very greatly reduced.

One more disturbing factor affecting olfactometry should perhaps be mentioned — the effect of concurrent impressions from other senses. All the senses affect each other. High and low tones are said to increase olfactory acuity at least to some odours and visual stimulation sometimes increases, sometimes decreases it. And, as long ago as 1833, Stark produced evidence that the extent to which some substances (cotton, silk, wool) absorb odour varies with their colour.

After olfactometric measurements have been made, how are the results to be expressed? It is perhaps possible that a low threshold value does not necessarily correspond to high intensity but it is quite generally assumed that this is the case and concentration of odorant at the threshold has usually been given in grammes or gramme-molecules per litre, in milligrammes per cubic metre, in parts per million, per billion, per milliard or in a similar way. However, the suggestion has been made that results should be expressed more correctly as thermodynamic activities (the ratio of the partial to the saturation pressure) and Laffort, who recently examined the values given by 24 authors for 192 pure substances (some inorganic substances included) during the period 1851 to 1963, proposes to measure the "olfactory potential" or *p.* ol. (corresponding to *p*H or *p*K). His values range from about 4.0 for butane to about 12 for a synthetic musk (2 : 4 : 6-trinitro-*tert*butyltoluene), high potential corresponding to low minimum perceptible. He further affirms that odorous power is a function, not only of molecular volume and chemical constitution but also of coefficient of partition between air and water (*i. e.*, aqueous fluid on the receptor) and he has given a formula, involving *p.* ol., from which threshold values can be

[5] MacLeod, however, maintains that radio-active tracers, like chemical analysis, chromatography and spectro-photometry, is not nearly sensitive enough. "The Geiger counter", he says, "is an instrument much less sensitive than our sense of smell." This sense depends on an adsorption process with stereo-selection and all theories of olfaction worth considering are based on the *shape* of the molecules of the odorants concerned. Amoore's theory (see p. 166) is not credible. — He simplifies unjustifiably.

calculated (in the aliphatic series, at least), given the experimentally determined value for one member of the series.

But whatever method of expressing results is adopted, we cannot cease to be astonished at the almost unbelievably minute amounts of material that can be detected by smell. Although, as Von Skramlik has calculated, the number of molecules of odorant (about 55 are here involved) present in unit volume at the threshold ranges from 10^8 to 10^{17}, it also seems to be true, according to evidence adduced by Stuiver, that one molecule of some odorants is enough to activate one sensory cell and that a sensation may be perceived if as few as forty cells are activated simultaneously.

In this connection, reference may be made to an ingenious and unusual procedure developed by Devaux for estimating the actual amount of material emitted by an odorous source.

Emission of Perfume by Flowers

> Solo chi sa cogliere ed apprezzare con riguardo e devozione i lati ai più nascosti delle cose e dei sentimenti e le piccole gioie che la vita ci offre, solo costui può relativamente risolvere il problema della propria felicità, proprio così come la vera essenza odorosa che solleva il morale e carezza i sensi non è quella costruita dall'uomo con aspre ed aridamente appariscenti combinazioni chimichi ma bensi quella che proviene dal semplice ed umile fiore amorevolmente coltivato, pazientemente raccolto e religiosamente trattato.
>
> *Guglielminotti*

Devaux also made photographic and even cinematographic records of the process of perfume emission. After moistening a mercury surface, he at once covered it with a thin coating of powdered talc and fixed over it, at a distance of a few millimetres, a petal of a flower, this petal being fastened to the under side of a glass plate. During the following few minutes he observed that the talc coating below the petal was slowly but regularly displaced in such a way that a "stain" with sharply defined edges was produced. This stain marked the area where the odorous vapour from the petal had been adsorbed by the mercury. Odourless petals produced little or no stain, any effect which they had being due to the emission of water vapour. But flowers such as the rose, lily, jasmine and carnation produced stains which, Devaux believed, are monomolecular layers. They can be smelled, he says, by a keen nose or even, if the layer is caused to shrink so that it ceases to be monomolecular, by a less practised nose. If the breath is exhaled on the stain, a "two-dimensional fog" is formed on the stain-area and this fog, which is similar to that produced by breathing on glass, persists for a long time if the surface is cold. If it be assumed that the molecular weight of the odorous matter lies between 120 and 200 and that its density is 0.9, then the weight of one square centimetre of the monomolecular layer of rose perfume is 0.7×10^{-7} g., this being the amount emitted in a few minutes. Hence, in twenty-four hours, each gramme of rose petals emits a few milligrammes of perfume. Petals of jasmine emit 10 mg in the same period.

The accuracy of these very interesting results appears to depend, it may be added, on at least two assumptions not mentioned by Devaux. It is assumed, in the first place, that all the perfume from the petal is adsorbed by the mercury surface. And, in the second place, it is assumed that all, or almost all of the adsorbed material is, in fact, perfume.

Perhaps it should be mentioned here that it may be necessary, for olfactometric purposes, to know the vapour pressure of an odorant although it may often be impossible to determine this pressure directly, the value being so low. Because this is so, Chamuleau and Noyons (1932) developed a spectrographic procedure which yields satisfactory results indirectly. And it is also interesting to record that, in 1928, Tammann and Oelsen, proceeding in the opposite direction, used their sense of smell to determine the vapour pressure of butyric acid and camphor.

Theory

Le problème de l'odeur est un problème particulièrement irritant pour un esprit scientifique. À l'imprécision du vocabulaire, s'ajoute le manque de bases solides et d'expériences positives. Les données précises sont encore très rares. On ne saurait s'en étonner, étant donné la difficulté du sujet.

Bornand and Martinet

Pour le moment, il nous faut simplement constater que nous savons peut-etre beaucoup de choses, mais que nous en comprenons peu. *Beets*

We should not be surprised either that there have been and still are many theories of olfaction. Some, like the almost occult theory of Leideritz (1959) which seems to consists of a mixture of psychology and superstition concerning human and animal attractions and repulsions and the theory of Juhász (1926) who maintained that the most important characteristic of an odour is its "pitch" which is the sole basis for an orderly systems of odour classification, seem to have little or no solid foundation. Perhaps no more firmly based are the relativity theories of Gattefossé, the distinguished perfumery expert, and Guglielminotti. Gattefossé's paper, *Einstein et les Parfums* (1922) suggests that the various conflicting theories of olfaction may be harmonized by taking advantage of the results of Einstein's calculations and experiments. A similar idea is the basis of the Italian author's *Influenza del trascendentalismo relativista sulle teorie chimico-fisiche dell'olfazione* (1925) Gattefossé, who appears to accept, in part, the views of Teudt (these are discussed later), speaks of alternate de-materialization and rematerialization, of electronic hypotheses and of special sub-atomic properties of odorous substances and Guglielminotti, in his speculations, tends to write as a poet and mystic rather than as a man of science. No doubt there is no harm in this but his argument, like that of Gattefossé, is not convincing however interesting. In particular, it is not at all clearly shown how, in respect of relativity and electron theory, odorous compounds should differ from non-odorous. And although much has been learned about relativity since 1925 it seems unlikely that the new knowledge can lend support to the opinions of these two speculators.

At various times it has been claimed that there exists some correlation between the electrical conductivity or the ionization of air and the presence in it of odorous material and experiments in support of this claim have been described. Thus, as early as 1908, Bordier and Nogier claimed that air and other gases *(e. g.*, nitrogen, carbon dioxide) when exposed to irradiation by a quartz-mercury vapour lamp acquire an odour not due to ozone or to oxides of nitrogen and that this odour can be removed by de-ionization. A few years later, Dubuisson advanced the opinion that odorants in the state of gas or vapour form ions that stimulate the sense of smell and Zwaardemaker and others, investigating the colloidal state, diamagnetism and what was termed "spray" electricity or "waterfall" electricity, likewise believed that they had disclosed a relationship between odour and electric charge.

Many years later, Erb revived this hypothesis in a somewhat different form. "Odor", he stated, "is the result of this electronic potential, differences in potential producing differences in odor. Odor is a measure of the speed and concentration of the electronic activity of an oxidation-reduction reaction of the odorophore." And, soon afterwards, Niccolini claimed that Erb's views confirmed his own theory which is that odour is perceived as a result of the energy due to deficiency of valency electrons. It seems clear, however, from the work of Guillot who used fully insulated instruments and prevented the adsorption of moisture on them, that odorants in the air do not produce any special electric charge. Working with Madame Guillot, he examined the electrical behaviour of air containing citral, camphor, vanillin, menthol or skatole but was unable to observe any increase in conductivity resulting from the presence of these odorous substances which, moreover, did not produce any condensation in a cloud chamber. As regards ionization of the air, presence of ozone, or existence of condensation centres of water vapour, he concluded that the presence of odorous organic vapour in air does not confer any special electrical property on it.

Another theory that was subsequently shown to be ill founded was Tyndall's heat absorption theory. Tyndall found, in 1865, that air rendered odorous with aniseed or other essential oil absorbed from 30 to 370 times as much radiant heat as did dry odourless air and, about seventy years later, Castex, who believed that the emission of odour from flowers is accompanied by an increase in temperature, affirmed that there is a relationship between odour and heat production (and also between odour and electric charge and absorption spectrum). However, as Grijns was able to show in 1919 when examining a large number of odorants, Tyndall's apparaturs must have been in some way inadequate or defective and, in fact, there is no relationship between intensity of odour and absorption of radiant heat.

Because of the predominant biological importance of oxygen, it is natural that this element has sometimes been credited with special olfactory properties. At the beginning of the present century, Sakharov (1901—02) passed hydrogen into the nose until (as he believed) all oxygen was removed and then added odorant (aniseed, nitrobenzene) to the hydrogen stream. He found that no odour could be perceived until oxygen (air) had been admitted (*e. g.*, during exhalation). Later, he added that, in addition to hydrogen, a substance that removes oxygen must also be used. Sakharov's apparatus and procedure were probably somewhat primitive and experiments carried out much later are stated to have shown that air can be replaced by odourless gases (argon, carbon dioxide, helium, hydrogen, nitrogen) without altering ability to smell and hence that oxygen plays no specific part in olfaction. (It may be doubted, however, whether even in the modern experiments, *all* the oxygen was replaced by the other gases.)

Enzymes

In view of the important part played by catalysts and enzymes in many biological processes, it is not surprising that some investigators believe they (particularly the oxidases and phosphatases) may also be involved in the mechanism of olfaction Thus, Alexander (1951) has suggested that odorants may affect the enzyme (catalyst) balance of the olfactory cells by modifying existing enzymes, by forming new enzymes (neo-catalysts), by inhibiting all or part of the activity of normal cellular enzymes or in all these ways and Kistiakowsky (1950) has pointed out that odorants may be assumed to act selectively by inhibiting chains of enzymically catalysed reactions, each step in the chains being catalysed by a different enzyme. No basic physical principle is infringed by making such an assumption which also takes account of or explains very great sensitivity to minute concentrations of odorous substance, correlation of olfactory intensity with extent of inhibition, change of odour quality when concentration changes, reversible and irreversible effects, and so on.

The inhibitory and stimulatory action of odorants (*e. g.*, vanillin) on enzymes, especially on alkaline phosphatase, has been etamined by Baradi and Bourne (1948, 1951) who also studied the distribution and concentration of the enzyme in the olfactory mucous membrane of cats and rabbits. Alkaline phosphatase, apparently, is present in relatively high concentration in the olfactory mucosa of man and animals. It has also been suggested (by Lauffer) that choline esterase, because of its biological importance, is especially likely to be involved in the olfactory process and Niccolini (1932—33,1934—35) concluded as a result of his study of the odour of coumarin and its derivatives that oxidases may likewise play a part, enzyme action possibly accounting for the rapidity and reversibility of the olfactory process. Possibly also, his contention that the concentration of thiocyanate (which is said to "mobilize" oxygen) in the nasal mucus varies with olfactory acuity is in some way connected with the activity of enzymes.

If enzymes indeed play a decisive part in olfaction, it could easily be understood that substances having quite different chemical structures could have very similar odours since they might inhibit the action of the same enzymes. Enzymes, however, are stereo-specific and Wright (1963) has pointed out that it is most unlikely that they could be affected alike by both stereo-isomerides. For if the olfactory qualities and intensities of pairs of such isomerides differ at all, the differences are never other than slight. Accordingly, he holds that theories of retardation or acceleration by enzymes as the first step in odour perception must be rejected as should any theory that this step consists in the enzymic transformation of the odorants. The primary process is probably *physical* not chemical or enzymic.

Whether or not this is so, it is of interest to note that Bronstein, working with the olfactory linings of guinea-pigs, rabbits and tortoises, found some evidence that odorants (camphor, thymol) cause the thiol groups in the linings to increase and possibly affect the nucleic acid contents also. In this connection, reference should be made to the suggestion, advanced chiefly by Italian investigators, that the nasal secretion may contain substances that "activate" odourless material rendering it odorous, or convert weak odorants into stronger ones by affecting their electro-chemical character in some way. This suggestion is based on the fact that anosmic patients who could not perceive the odours of coffee, ionone, rhodinol, skatole and vanillin when these were inhaled in the normal manner, could perceive them if they were first drawn into the nasal cavities of healthy persons.

The part played in olfaction by the olfactory area has, of course, been frequently considered. About 45 years ago (1920) in his peculiarly interesting paper entitled *Die Grundlagen der Geruchschemie*, Ruzicka advanced his "osmoceptor" theory to which he referred with approval at the Society of Chemical Industry symposium on "Molecular Structure and Organoleptic Quality" in 1957 and which still provides an acceptably satisfactory explanation of at least some olfactory phenomena. Because he was not convinced that the so-called odours of inorganic compounds really *are* odours, Ruzicka decided to consider organic compounds only. Turning to the problem of classification and surveying the chemistry of these compounds, he quickly concluded that an attempt to classify on a chemical basis could not be successful. Chemical and physical considerations, however, led him to suppose that the properties most essential for ability to stimulate the olfactory nerves are specific solubility in air (in Erdmann's sense) and power to enter into chemical reaction with certain constituents of the olfactory mucous membrane. These constituents, he named *osmoceptors.* Odorous substances, of course, must be sufficiently volatile to pass into the air and thence into the osmoceptors when the air is drawn over the membrane and, in view of this, the extent to which the substance, solid, liquid on in solution volatilises and is then distributed between the air and the osmoceptors is of fundamental importance. Very important also, in Ruzicka's opinion, is his supposition that there are two kinds of osmoceptor — primary and secondary. An odorous substance, on reaching the olfactory mucous membrane, combines first with primary osmoceptors only. If there are sufficient primary osmoceptors to combine with the quantity of odorous substance present, or if such osmoceptors can be reformed quickly enough by the organism, then the quality of the odour remains constant. But if all the primary osmoceptors are used up, the substance may then react with secondary osmoceptors and consequently the quality of the odour may change. The change will occur slowly if the concentration of odorous substance is low, rapidly if it is high. In some cases the change may occur so rapidly that a "mixed" odour is at once produced. An explanation is thus provided of the difference sometimes observed between the odours of the concentrated and dilute forms of the same substance. If a substance reacts with the primary but not with the secondary osmoceptors, odourlessness (fatigue) follows saturation of the primary osmoceptors and lasts until new ones are available. Fatigue for one odorous substance will presumably extend to substances having similar or identical "osmophoric" groups but not to other substances [6].

Although Ruzicka's paper is most interesting and most suggestive because of its account of the osmoceptor theory, it also deals with other aspects of the question. The relationship between chemical constitution, physiological action in general and olfactory stimulation in particular, is discussed in some detail, account being taken of the various physiological and psychological factors involved. It has been objected that the theory is too speculative and it must be admitted that the existence of the two types of osmoceptors has never been directly demonstrated. Further, Henning has rightly remarked that Ruzicka deals solely with peripheral processes although it may be that central processes play a predominant part Nevertheless, wise speculation seems justified so long as experimental evidence can be but partially acquired. Moreover, the distinguished American chemist Sumner, in 1953,

[6] It is, perhaps, not entirely fanciful to suppose that Gesteland's very recently (1966) advanced theory that each olfactory receptor acts like a "complete nose", capable of detecting some differences between some odours is an extension of Ruzicka's hypothesis.

put forward a somewhat similar theory (although "against his better judgment"). He considered that the energy of olfactory stimulation may be derived from the breakdown of adenosine triphosphate.

Very soon after Ruzicka published his paper, another Swiss author, Tschirch, asked himself the question: What, then, are the conditions which a substance must fulfil in order to have an odour? The first condition, he felt, is solubility in air, solubility, that is, in Erdmann's sense. Volatility and boiling point are of secondary importance. On the other hand, the air being regarded as a solvent, the distribution law and presumably also the phase rule are of first importance since they show why the odour of a dissolved substance may be stronger than that of an equal amount of the substance in the undissolved state and why a dilute solution may have an odour stronger than that of a concentrated solution of the same substance. The solution of the odorous substance in air is assumed to be molecular, not colloidal. Tyndall's alleged demonstration that the capacity of air to absorb radiant heat is increased, often enormously, by minute quantities of odorous substances and Berthelot's claim that camphor does not pass into the Torricellian vacuum are cited in support of the view that air dissolves these substances. It is possible, if not indeed probable that the substances which are perceived by the olfactory nerve are not the odorous substances themselves but "fragments" detached at the very high dilutions at which perception can occur. Various grounds are given for supposing that such fragmentation may take place. It is known, for example, that there is often a qualitative difference between the odour of a substance and that of its dilute solution. Also, the similarity in odour of substances of quite different constitution strengthens the supposition. There is a possibility, too, that the oxygen of the atmosphere plays some part in the fragmentation. However that may be, it is as a result of the coming into contact of the material with the receptors of the olfactory mucous membrane that sensation is produced. Perhaps the first stage of contact consists in dissolution in the aqueous-lipoid material on the surface of the membrane. In any case, it is assumed that new labile compounds or complexes are produced by the immediate interaction of colloidal plasma of the membrane with the odorous substance. These compounds or complexes stimulate the olfactory nerve and hence they are the true odorous materials. The substances that have heretofore been called odorous have really no odour: they acquire it by reacting with the plasma. Conversely, substances (or fragments) which, although soluble in air, do not react with the plasma cannot arouse olfactory sensation. This holds also, in some measure at least, for very reactive substances, such as acid chlorides, which attack and destroy the plasma. Olfactory fatigue is explained by supposing that the plasma may become saturated with odorous material. Only when the complexes decompose does the fatigue pass away. Moreover, since an organ fatigued by one odorous substance can still perceive others, it is supposed that molecules of the plasma are attacked at different points by different substances. Great individual variations in olfactory acuity are encountered because the degree of organization of the olfactry plasma varies greatly from individual to individual. Similarly, animals differ from man, from each other, and from species to species. There is no evidence that specific olfactory nerve fibres or groups of such fibres perceive particular odours only and nothing is known of a "localization" of odours in the olfactory region. The extraordinary complexity of the protein molecule with its many points open to attack suffices to explain the multiplicity of olfactory sensations and to account for olfactory compensation. Many conditions must be fulfilled, said Tschirch, before a substance can emit an odour. It must be present in appropriate physical form at an appropriate temperature. It must be sufficiently soluble in air. And, finally, it must be capable of reacting with the olfactory plasma to form a labile complex that will stimulate the olfactory nerve. If a single one of these conditions is not fulfilled, there will be no odour.

It is obvious today that some of Tschirch's opinions cannot be accepted. But it is also far from certain that his theory does not provide some valuable suggestions concerning the mechanism of olfaction.

In England, at the time when the Swiss authors wrote, Durrans proposed his "residual affinity" theory in a long paper in the *Perfumery and Essential Oil Record.* His idea was that the odour of a compound was associated with the extent to which it was chemically unsaturated or possessed unsatisfied valencies. Many objections to this view at once spring to mind and, in fact the hypothesis is now of historical interest only.

Since the olfactory receptors are always covered with an aqueous fluid, it is natural to suppose that substances which cause the sensation of smell must first dissolve in this fluid. As long ago as 1884, Aronsohn had affirmed that when odorants are dissolved in sodium chloride solution and injected into the nose, the odours *are* perceived and also that odours inhaled through the mouth and then exhaled through the nose are likewise perceived. About thirty years later, Backman, having conducted experiments with amyl salicylate, camphor, menthol, piperonal and other odorants, advanced his theory that intensity of odour increases as solubility in lipoids increases and solubility in water decreases. Thus, as the distribution coefficient for lipoid : water increases, so does the intensity of odour. He indicated, however, that his hypothesis applied to homologues and isomerides only.

One of the authors who supported the theory was the Spanish investigator Martin-Calderin. Another was Sass who has recently (1952) claimed that solubility in lipoids is the factor governing the intensity of odour. The solubility, Sass states, increases with increase in similarity of the structure of the odorant to that of lipoid. The intensity is also proportional to volatility and hence "total" intensity is the product of the two. Lipoid-solubility, moreover, increases very roughly, with increase in molecular weight. In accordance with his views on this subject, Sass distinguishes four classes of odorants, namely, hypo-odorants, those that are very volatile; general odorants, those that have moderate volatility and lipoid-solubility; hyper-odorants, those that are very soluble in lipoids; and trans-odorants, those that cannot be perceived by human beings.

An unusually interesting theory which in some ways resembles that of Backman, was evolved by Ehrensvärd during the second world war. He found that when a physiologically active organic compound is introduced, at the appropriate point, into an electric cell in which an oil is in contact with an aqueous solution of an inorganic salt, a change takes place in the electromotive force that normally exists at the interface. Since the amount of material required to produce a measurable alteration in the force is often very minute and since the properties of such a "model" cell system may be regarded as being in some degree analogous to those of the heterogeneous constituents of the protoplasm of living tissues, it is not unreasonable to suppose that a relation may exist between physiological activity and power to change the electromotive force. More particularly, it may be thought that, in the primary stage of perception, the molecules of odorous substance which come into contact with the bio-electric system of the olfactory mucous membrane produce their effects by altering the electromotive forces existing at phase boundaries within the system. This is the supposition on which one of the most modern and most exhaustive theories of the first stages of olfactory perception is based. Ehrensvärd set out to investigate the physico-chemical basis of smell and taste but it soon became clear to him that, in fact, he must first study the action of organic compounds on physiological systems in general. Accordingly he made detailed investigations of the structure of the oil-water interface and of the differences in electrical potential that are there set up. Then he investigated the effects of inorganic and organic compounds on both the structure and the potential differences. Moreover, he greatly increased the value of his work by applying to it the knowledge gained in recent years of the inner structure of organic molecules and of their electronic configurations. He pointed out that, at the interface, the changes in potential difference produced by such molecules would be affected by adsorption, by distribution between oil and water, by diffusion, by electromeric and inductive displacements of the electrons of the organic compounds producing the changes and by the spatial structure of these compounds.

In the field of general pharmacological action, it has not yet been found possible to correlate biological activity and change in phase-boundary potential although some degree of parallelism has been demonstrated in the case of the toxicities of compounds in honologous series. Accordingly, it is not to be expected that Ehrensvärd's studies would solve the problems of olfaction regarded as a special case of chemical action on a physiological system. Nevertheless, they enabled him to suggest a classification of odours based on the significance of reactive groups and spatial structure and they made it quite clear that if there be, in fact, phase boundaries in the receptor cells of the olfactory mucous membrane, the minute quantities of odorous substance which may be supposed to cause sensation are quite large enough to provide an adequate stimulus by changing the local potential at such boundaries.

Ten years in advance of Ehrensvärd, the Japanese investigator Hukui conducted some elegant experiments in which the lowering of surface tension by odorants at an air-water interface was examined. He caused a current of odorous air to pass over one side of a small stick suspended in such a way that it just floated on the surface of pure water. So long as the surface activities on the two sides were unequal, the stick moved one way or the other. Hukui supposed that similar movement of the hairlets in the olfactory mucosa occurs, so initiating the olfactory process.

In the studies undertaken by Ehrensvärd, the systems studied may be regarded as having a somewhat distant resemblance to a model of the biological system concerned. Considerably closer to nature is the colloidal model of Bungenberg de Jong and Saubert, described in 1937. However complex the olfactory process may be, these workers say, it seems obvious that, in the initial stage, a reversible change is produced in the olfactory cells as a result of penetration by gaseous molecules of odorous substance. Possibly the change is in permeability. On general physiological grounds, it is believed that the constituents of the cells chiefly affected are the phosphatides since these, unlike carbohydrates, proteins and so on, are very sensitive to the action of organic non-electrolytes even in very low concentration and are also both hydrophilic and lipophilic. For technical reasons, Bungenberg de Jong and his collaborator studied, not phosphatides, but oleates which behave very much in the same way. The emulsified oleates, in the form of minute droplets, known as coacervates, were examined microscopically. It was found that vapours of some organic substances cause "shrinking" when the concentration is low, but "swelling" when it is high. When the vapours pass out of the coacervates, restoration of the original volume occurs.

It can easily be seen how certain olfactory phenomena can be correlated with these findings. Thus, compensation would occur when the shrinking produced by one odorous substance is counteracted by the swelling produced by another acting simultaneously. Again, the change in odour quality often observed in homologous series is paralleled by a change from swelling to shrinking. (Methyl, ethyl and propyl alcohol cause swelling but butyl, amyl, hexyl and heptyl alcohol cause shrinking.) Substances that cause shrinking when in low concentration but swelling when in high may be those that have one odour quality when dilute and another when concentrated.

Probably the most recent of the biological models, the "cell-puncturing" model, is that of Davies and his co-workers. They suggest that the red blood cell can serve as a model of the olfactory membrane and have shown that odorants (butyric acid, ionone, musk, octanol, skatole, *etc.*) accelerate haemolysis by saponin in proportion to their odorous strength (*i. e.*, their threshold values). The molecules of odorants are said to be bulky, rather rigid and awkward-shaped. They puncture the cell walls, acting like a trigger which produces effects that are then very greatly amplified so that potassium rapidly passes out and is replaced by sodium. Of powerful odorants, perhaps only one molecule per cell is required and it is significant that the number of molecules present at the threshold, about 10^8, is also, approximately, the number of olfactory cells in man. Of less powerful odorants, more molecules must penetrate at the same point. Hence more are required and intensity of odour is lower. Threshold values depend, quantitatively, on the adsorption energies of the odorants passing from the air to the lipoid-aqueous surfaces of the olfactory membranes and on the sizes and shapes of the molecules of the odorants. Since adsorption energies can be calculated from vapour pressure and interfacial tension measurements, the olfactory threshold for any substance may be predicted by applying an appropriate mathematical equation.

It is obvious that this is a theory for which much is claimed and, in fact, its authors affirm that there is very good correlation between calculated olfactory thresholds and observed values. Probably those who are familiar with theories of olfaction, however, will reserve their judgement [7].

The triggering effect said by Davies to be produced by cell-puncturing, is attributed by Dravnieks to adsorption of odorants on ferro-electric compounds (spontaneously polarized "molecular dipoles"). It is quite possible, he says, that these ferro-electrics, which may be in a state of strain before the adsorption occurs, have a triggering action and, serving as transducers, may provide the degree of amplification which may be assumed to be necessary in order to yield the quantum of energy required. For the extremely minute quantities of odorant that stimulate the olfactory receptors seem to be totally incapable, by themselves, of producing their effects without some such amplification.

Adsorption

Whatever the fate of Dravnieks' theory, his view that the initial step in the olfactory process is adsorption of the odorant on the appropriate

[7] "We conclude", wrote Davies and Theimer recently, "that the odor type is a purely physical property of the odorant molecules determined quantitatively by their rate of desorption, molecular cross-sectional areas and the dimension ratio *L/B* [length/breadth]." This conclusion is drawn from a study of more than fifty compounds having musk odour.

receptors is very widely accepted. The very ready adsorption of odorants on metallic and other inorganic surfaces can be a serious obstacle to precision in olfactometry. But organic matter, including living tissue, also readily adsorbs odorous (and other) compounds and there is strong evidence that this adsorption plays an important, perhaps an all-important part in the mechanism of olfaction. Thus, Watson, one of the few early British investigators to study the subject, found that there is a correlation between intensity of odour and adsorption from aqueous solution by charcoal. (Like Hukui, he also observed a correlation between this intensity and lowering of surface tension of aqueous solution.) Later, the Russian worker Ouchakov conducted a very large number of adsorption experiments using carbon, cork, glass, rubber and other adsorbents with odorants such as benzaldehyde, menthol and vanillin. He obtained some indications that, with one and the same adsorbent, the extent of adsorption increases with the molecular weight of the adsorbed substance. More important, he found that both adsorbents and adsorbed substances exhibit selective power. Ouchakov's work (published in 1930) seems to have been neglected by subsequent investigators but since the second world war the subject has been closely studied on various occasions.

In 1949, Dervichian reviewed the subject admirably and, in 1955 Jones, attempting to disclose a correlation between olfactory thresholds and various physical properties, examined twenty pure chemicals which were smelled by 84 observers and concluded that "the basic process in olfactory stimulation is adsorption on the receptor surface". About the same time, Moncrieff conducted experiments in which the adsorption of various odours on inorganic and organic material (*e. g.*, the nasal passages of decapitated sheep) was studied. Although these experiments are not without value, it has to be pointed out that, because of the complex nature of the tissues concerned and because some, at least, of the odorants used (*e. g.*, benzaldehyde) are chemically very reactive, processes other than adsorption must very probably have accounted for the loss of odour which occurred when odorized air traversed the passages. Moreover, no steps were taken to discover whether odourless substances were also adsorbed. Giving reasons why a physical theory (adsorption) is to be preferred to a chemical hypothesis, Moncrieff pointed out that an olfactory chemical reaction would have to be reversible and very fast and that the reaction products would have to be very rapidly decomposed or removed.

Another worker who has recently sought for a correlation between physical properties and odour is Schutz. He has claimed but has produced little evidence to show that by study of these properties accurate prediction of strength and pleasantness of odour is possible in many cases. And he has stated that substances with strong properties of adhering to surfaces usually have weak odours.

Despite all that has been said above about the olfactory significance of adsorption, it must be admitted that odourless substances also appear to be adsorbed in the same, or almost the same way and perhaps to the same extent as are odorants. So it is not surprising that Le Magnen has expressed himself as follows: "On considère généralement comme évident que les corps odorants possèdent une aptitude particulière à s'adsorber sur les surfaces quelles qu'elles soient. L'évidence n'est qu'apparente et tient simplement au fait que l'adsorption des substances odorantes sur les surfaces, précisément parce qu'elles sont odorantes est directement et empiriquement sensible tandis que la fixation sur ces mêmes substrats des corps inodores n'est révélée qu'à l'aide de techniques de mesure délicates."

Although it would be difficult to refute Le Magnen's contention, it has seemed possible to some investigators that one of the modern theories involving adsorption, namely the "site-filling" theory might explain olfactory phenomena. This theory which had often been advanced in embryonic form, was revived by Moncrieff nearly twenty years ago and has now been elaborated and greatly extended by Amoore who has gone to great lengths in his attempts to establish it. The olfactory region, he asserts, contains seven perfectly distinct receptor site of definite shapes and dimensions. Substances with molecules so constructed that they will fit more or less closely into one or other of these sites will have one or other of the "seven primary odours" which he claims to have identified. These odours are: camphoraceous, pungent, ethereal, floral, peppermint, musk, putrid. Several quite well designed experimental tests of this theory have now been made, chiefly by Johnston and his co-workers in the U.S.A. and it has been claimed that the results give it a large measure of support. They even show, it is said, that predictions based on the theory have been fulfilled. Some of those who find the theory acceptable have admitted that it requires modification, however.

But there are many objections to it. It is based very largely on conjecture, speculation and subjective impressions and the quality of some of the evidence adduced in its favour is very poor. In particular, it must be regarded as extremely astonishing that no anatomical or physiological evidence can be given concerning the actual sites which the odorant molecules are believed to occupy [8].

Vibrations

In spite of the very obvious objections to them, non-material theories of olfaction have frequently been advanced. They have been advanced, of course, because of the many difficulties, real or apparent, encountered in

[8] Amoore has now, apparently, withdrawn much of his theory and has explicitly identified another primary odour, namely, the sweaty odour of *iso*butyric acid.

explaining many olfactory phenomena on a material basis, difficulties that have led to statements such as that of Charles that "perfume is revealed to the sense of smell by means of vibratory waves", a theory favoured by Ramsay but dismissed as "bosh" by Oliver Lodge. The evidence in favour of the material theory, however, is overwhelming. But it nevertheless seems worth while to give an account of some of the vibrational theories which have followed the brief suggestion made by Wood in 1892. And, of course it must be borne in mind that some of the vibratory theories do not differ entirely from material theories since they require that contact between odorous substance and olfactory receptor must take place.

Of particular interest is the theory of Teudt who appears to have been the most vigorous protagonist of this view. Writing in the early days of the electron theory (1913—1920) he explained his position in several of the scientific and technical journals of his country *(Biologisches Zentralblatt, Deutsche Essigindustrie, Prometheus, Wochenschrift für Brauerei, Zeitschrift für anorganische und allgemeine Chemie)*. Odour, he held, is a molecular, not an atomic property. The valency electrons which are shared by the atoms constituting the molecule of an element or compound (but not the other electrons) are the agents which stimulate the olfactory nerve. They do this because, being alternately attracted and repelled by the atoms which share them, they acquire a special rhythmic motion which is not acquired by the other electrons. The period of vibration varies from molecule to molecule and is, of course, affected by the nature of the molecule. In the olfactory nerve are molecules of various kinds which likewise possess vibrating electrons but the vibrations there are not, by themselves, strong enough to cause olfactory sensation. When, however, the vibrations in an odorous substance have a period corresponding to that of any of the electrons in the nerve, resonance takes place and odour is perceived, the action being somewhat analogous to the electrical production of an induced current. Accordingly, it is not necessary for the substance to come into contact with the olfactory nerve-endings. Moreover, since odorous substances induce vibrations in non-odorous molecules (the same non-odorous molecule can acquire different odours according to circumstances) such as those of water in the atmosphere, sensation may be produced in the absence of the odorous substance provided that odourless molecules in which appropriate vibrations have been induced come sufficiently near to the olfactory region. Teudt claimed that his theory provided an *easy* explanation of *all* olfactory phenomena including, of course, those involved in the tracking of human beings by dogs as described by Romanes. Discussing the reasons why some molecules only which contain valency electrons are odorous, he stated that, in complex and heavy molecules, interference prevents production of appropriate vibrations. In certain simple molecules (those of methane and ethane, for example) which would be expected to have odour but have not, the symmetry causes the vibrations to be "balanced out". (Olfactory compensation is also due to balancing out.) Since odour appears the more readily, the more firmly the electrons in an element are attached to the nucleus, metals and their simple compounds have little or no odour. Further, the less metallic the character of a compound, the greater is the possibility that it will have an odour.

It is not surprising that Teudt has had many opponents for his theory is clearly very open to attack. His chief antagonist was Heller who replied also in *Biologisches Zentralblatt* and *Prometheus* as well as in *Deutsche Parfumerie-Zeitung*. Even more critical than Heller was Von Skramlik who roundly characterized the theory as pure phantasy, entirely lacking any experimental basis.

A vibratory hypothesis much more comprehensive than that of Teudt was put forward in 1933 by Krisch, encouraged by Max Planck. Krisch's theme is a new theory, not merely of olfaction, but of all the senses. He begins by suggesting that, in the last instance, all energy is of the same kind and by recalling that, early in animal development, there exists only a single undifferentiated sense organ with which all impressions derived from the environment are perceived. From this organ, the "five" sense organs were later evolved. Since this is the case, it is not unreasonable to suppose that all the sense organs respond to the same kind of stimulus, namely, electronic vibration. Thus, the receptors of the various sense organs are resonators of various magnitudes adapted to react to certain impulses. The internal structures of all the receptors are presumed to be similar but the receptors of each particular sense organ are so modified that they receive impulses of certain magnitudes only. The minute particles of odorous substance that approach the olfactory receptors, directly through the olfactory cleft or indirectly in the blood stream, radiate energy and so cause resonance in the receptors. Actual contact between odorous substance and receptor is not essential. A consequence of this supposition is that, since the characteristics of an odour depend on the grouping of electrons within the molecule, there must exist, for each particular odour, a specific series of impulses. Such impulses, arranged in magnitude and series, must constitute an "olfactory spectrum" in which certain magnitudes play a predominant part. Another consequence is that any resonator, no matter to which receptor it belongs, must be capable of reacting to impulses normally received only by resonators of *other* sense organs, provided only that suitable means of transmission are employed. Thus the energy derived from a lamp, a violin, a stove or a cheese may possibly, in accordance with the theory, be so influenced by human intervention that the violin "smells", the cheese "warms", the lamp "sounds" and the stove "shines". Krisch does not confine his hypothesis to the first stages of perception. He further suggests that the impressions acquired by the receptors may be transmitted, also by resonance, to the brain, where, by repeated resonance, minute particles of the tissue are oriented in the same direction. Association and memory are then to be explained by supposing that the impressed memory pictures rapidly and easily re-form when the same or similar sequences of impulses recur.

Writing at about the same period as did Teudt, Heyninx began by asking himself the question: What form does odorous energy take?, when he undertook his study of olfactory physiology. The mechanical, chemical and physical theories advanced by his predecessors failed to satisfy him as did four of the five physical hypothesis which he considered worthy of close examination. The fifth suggestion, that odorous energy is moleculo-vibratory and acts by direct contact with the olfactory apparatus, seemed to him the sole admissible view and on it he reached the following conclusions:

1. All odorous substances emit odour-bearing molecules which spread themselves throughout the surrounding medium. If this medium is air, these molecules are present as vapour whilst if it is water, they form very dilute solutions.

2. These molecules, which have specific odorous and ultraviolet vibratory motions of wave-lengths 0.35 μ to 0.20 μ can be arranged so as to form seven fundamental groups of odours, namely, the sharp, the rotten, the offensive, the burnt, the spicy, the vanilla, and the ethereal. A physical system of classifying odours is thus provided [9].

3. Having penetrated into the olfactory cleft, the odorous molecules are adsorbed on the surface of the olfactory mucous membrane where they form a layer having a thickness of about 0.13 μ.

4. The specific vibrations of the "odoriphoric" groups of the odorous molecules cause the olfactory mucous membrane and the cilia of the olfacto-sensory cells (Schultze's cells) in it to vibrate.

5. The olfactory pigment probably acts by olfactory resonance, so intensifying the moleculo-vibratory action of the odour-bearers.

6. The human olfactory "tessitura" has a range of at least an octave between the wave-length values of 0.35 μ and 0.20 μ.

Heyninx published these conclusions in 1919 and, on a later occasion (meeting of the Belgian Study Group on Oto-Neuro-Ophthalmology, Dec. 1932) repeated them without answering criticism voiced in the intervening period. He reported also that he had now constructed two olfactometers based on his ideas and would describe these instruments in a future communication. However it seems that no such communication was made. In any case, despite his assertion that his physical theory does not break down under critical analysis and his claim that this theory alone can explain

[9] For more than one author (e. g. Amoore, p. 166) the number seven seems to have had a special attraction!

various olfactory phenomena such as compensation, it appears certain that there is no parallelism between absorption of ultra-violet light and odour quality. Indeed, according to Von Skramlik, it is only in very special cases that members of the same class of odours have the same absorption bands.

In his opinion, too, Heyninx was not only mistaken; he led at least one other, Stefanini, astray! This Italian investigator, having had an attack of influenza, suffered from anosmia for a time and, accordingly, was able to make special observations on himself. (The anosmia may have been due to therapeutic inhalation of the vapours of guaiacol and thymol, rather than to the disease.) The observations are most interesting and valuable: not so the deductions drawn.

Stefanini believed that there should be a close analogy between vision and olfaction and that, just as there are primary colours, so there must be three or four primary odours by mixture of which all other odours could be produced. From examination of spectra recorded by Heyninx, but without any other experimental support, he went so far as to suggest that, by mixing acetone, menthol and cinnamic aldehyde all other odours might be produced. His paper is entitled: *Sulla teoria della percezione degli odori.*

The next vibrational theory to be considered came from North America. It was advanced by Ungerer and Stoddard and it is particularly interesting because of its attempt to reconcile, to some extent at least, material and vibrational hypotheses. The authors accept the view that odour is produced by the contact of the molecule of perfume with the olfactory nerve. But they hold that the olfactory stimulus does not result from a biochemical or biophysical reaction following the coming into contact of perfume and nerve cell but is rather a consequence of intra-molecular vibrations in the perfume. "Odor", they say, "is the response of the delicate olfactory nerves to these intra-molecular vibrations, vibrations which the most delicate instrument yet devised can hardly detect, much less record and measure. Here is the explanation of the observed differences in the odors of substances. Every pure substance has its own rate of vibration; this vibration may be such as to influence the olfactory nerve and produce the sensation of odor or it may be too low or too high to be detected just as certain ether vibrations are outside the range of eye and ear; vibrations within a certain range produce a certain odor while those of another range produce a different one." Since it is the dynamic condition *within* the molecule, its characteristic vibrations which influence or determine its properties, the presence or absence of power to stimulate the olfactory nerves is likewise determined and the compounding of perfumes is the compounding of substances in such a way that their vibrations "combine" to produce certain rates or frequencies just as this can be done in the case of the vibrations or waves of light or sound.

Two examples are given of the way in which this theory may explain observed facts. It is known, for example, that the odour of a perfume may be profoundly modified by the addition of a very small amount of a substance having but a feeble odour. If a purely material theory of olfaction is adopted, it is very difficult to explain this. If, on the other hand, Ungerer and Stoddard's theory is adopted it is easy to conceive that the vibrations of the added molecules may greatly affect those of the perfume, so producing the observed change. Again, the puzzling loss of odour suffered by some odorants on purification (see above in the section on chemistry) is explained as resulting from the "combination" of vibrations. Such combination can equally readily be supposed to alter the quality or strength of the odour of one or more of the components of a mixture. (It may be added here that, if it were the case that odour arises only when two or more substances are mixed, then it should be possible to produce an odorous mixture by mixing two or more odourless substances. This, of course, has never been done.)

Ungerer and Stoddard's theory is largely, if not entirely, speculative but it does not lack its attractive aspects and it provides explanations for some difficult problems, at least. Durrans believed that it was much sounder than was Teudt's theory. But he rejected it, claiming that it failed to account for certain well known phenomena such as fatigue.

After the publication of this "theory of odour values", a long interval elapsed before another vibrational hypothesis was suggested. It came from South America where Castex and Vierheller discussed at length the absorption of light and of infra-red radiation by a large number of natural and synthetic odorants. In 1947, however, after Duranton had written his rather fanciful account of radio-biological vibrations and their relation to olfaction, Miles and Beck gave a report on the relation between olfaction and infra-red absorption at a meeting of the U.S. National Academy of Sciences. They found that bees seemed to be able to detect the presence of honey in a sealed box when this was provided with a window which allowed the passage of infra-red radiation but not when a window which prevented transmission of the radiation was used. Apparently, similar experiments with cockroaches and oil of cloves gave the same result. According to the theory which these fiindings are believed to support, "olfaction results when the radiation of heat from an organism's receptors is accelerated. The source of heat is the organism itself. The organismic source defines the critical region of the infra-red spectrum as that in the neighbourhood of the maximum of a black body at approximately 290—313° K. The receptors, being of the dimensions of the wave lengths in the region of this maximum (8—14 μ) can, by virtue of their size and shape, radiate selectively. Any substance having an absorption band in this region and coming within the radiation field of these receptors will cause them transiently to lose energy. This transient loss of energy is presumably the initial event in the process of stimulation of the olfactory receptors. Differential radiation losses of the receptors account for the olfactory quality as spatial stimulation." Beck and Miles claimed that their theory provided a (presumably non-sensory) means of distinguishing between odorous and inodorous substances, explained adaptation, accounted for differences between micro- and macro-osmatic animals, explained anomalies in chemical theories and provided a basis for the reinterpretation of some of the classic work in olfaction. "Further work", they said, "is expected to establish a small spectrum in which the four thousand to six thousand odors detected by the human nose can be plotted".

About ten years later, Narodny, having experimented with over 3000 bees, claimed that he had confirmed the findings of Beck and Miles. He stated, however, that the odorous attractant, honey, acts by causing resonance in the antennae of the insects so enabling them to distinguish flowers without from those with nectar — a very important matter, for the collection of one pound of honey requires that about 30,000 trips have to be made. Some support for Beck and Miles came also from Moncrieff who appeared to suggest that olfactory discriminating power in man may depend, in part, on the lengths of the olfactory hairlets and that these lengths are the same as those of the waves of infra-red radiation.

The amount of experimental evidence on which this theory is based seems very small in view of the magnitude of the theoretical claims and serious objections to it have been made. Young, Pletcher and Wright, for example, have pointed out that although butyl alcohol and its deuteroxyl counterpart have *different* infra-red spectra, their odours are *indistinguishable* and that both ordinary and heavy water are odourless although they have quite different spectra. Conversely, as Von Braun and his co-workers showed more than forty years ago, the odours of the *d*- and *l*-forms of some optically active compounds seem to be *different* although their infra-red spectra are *identical*. "These points", Young and his collaborators say, "would seem to offer an insoluble contradiction to any sort of infra-red radiation theories concerning (at least) the human olfactory sense".

Beck attempted, rather unsuccessfully, to refute these objections and subsequent experiments by Forrester and Parkins, Shkapenko, Gietz, Johnston and Robinson merely served to strengthen them. According to Robinson, both the infra-red absorption spectrum and the Raman spectrum of heavy water are quite different from those of ordinary water although, in harmony with "any radiation theory, heavy water should be extremely odoriferous since the absorption bands are strong and the vapour pressure is high".

One other vibration theory calls for comment. It is the molecular vibration theory of Dyson [10] recently modified by Hamilton Wright. Dyson held that "there is a general consensus of opinion that the only property

[10] In a recent (1965) review of Wright's excellent book, "The Science of Smell", Dyson has made many interesting comments and has supplemented his own earlier theory.

which can be correlated with odour is the periodic movement of the atoms or groups of which the molecules of odorous substances are built", and he affirmed that an odorous substance must possess "an appreciable vapour pressure, solubility in lipoid matter and one or more Raman shifts between 1500 and 3500 cm^{-1}". These values are the "osmic frequencies". According to Hamilton Wright, however, thermodynamic considerations indicate that, if the frequencies are to be active at body temperature, they must be below 700 cm^{-1} and he claims that his modification of the theory has "met all the experimental tests to which it has thus far [1957] been subjected. Huey Wright, however, who with his co-workers tested about fifty odorants, believes that the low-energy molecular properties of substances "do not appear to be *the* explanation for odor but they certainly appear to be correlated with odor to a statistically significant degree", and Hallam showed that at least one compound (ethyl carbamate) has the necessary spectral frequencies but has no trace of the postulated almond odour and is, indeed, almost odourless [11].

It is perhaps unlikely that waves of another kind might, in some way, affect odour. However, it is claimed in U.S. Patent No. 3,025,220 that perfumes can be modified and improved by being exposed to the action of sound waves of frequency of at least 20,000 cycles per second.

Beets has pointed out that not a few theories of olfaction are without value because the odorants with which they were concerned were impure, because the authors were mistaken or ignored necessary controls of for other fairly obvious reasons, and Foster's contention that many statements made by "authorities" are not sound must be accepted. It is perhaps not at all impossible that there is no dividing line between odorous and odourless, all substances being potentially odorous. If this is actually the case, any satisfactory theory must take account of it. Account should be taken, too, of the fact that distinctions between chemistry and physics, such as were suggested by Morimura and, later, by Middleton, are to a very large extent unreal in the olfactory context. Perhaps, also, the contentions of Herrmann, who conducted experiments with more than 200 substances, pure and mixed, including representatives of Zwaardemaker's and Henning's classes, and those of Achelis that odours are "complex" qualities not to be considered in isolation should be given more weight than is generally attributed to them. Herrmann's substances, in small flasks, were tested by psychologists and by psychologically untrained persons of both sexes and various ages, amongst them the investigator's four-year-old son and his playmates. (According to Henning, the concentrations of the odorous substances used by Herrmann

[11] Hamilton Wright is actively continuing his investigations and, as a result of further spectroscopic studies of simple derivatives of benzene (particularly cyanobenzene, nitrobenzene and benzaldehyde) has concluded that "the postulates of the vibrational theory of odour have been spelled out more precisely than ever before".

were much too high.) Henning's mistake, said Herrmann, was that he attempted to isolate olfactory sensation from all other sensations. His observers were enjoined to ignore, as far as possible, all memories, all associations, all non-olfactory qualities and to concentrate on purely olfactory impressions. They were required to abstract although it seems that they did not know from what they were abstracting! This procedure, Herrmann claimed, is entirely wrong. The whole organism is involved, not merely "a few nerves in the head". Odours have complex qualities and in their perception many factors, not only pleasure and displeasure, are included. Henning deliberately ignored all this. He became dogmatic and chose his classes arbitrarily, continued Herrmann. At one time, he stated that the molecular structure of an odorous substance was one of the properties indicating to which of his six fundamental olfactory groups it belonged and that the best confirmation of his psychological findings was thus provided. But, said Herrmann pungently, "an unproved psychological theory cannot be confirmed by chemical relationships!" As regards Herrmann's experiments, the results so far as they are recorded in the *Neue psychologische Studien* and in Schimmels *Bericht* are interesting but not very illuminating except from a narrowly psychological point of view.

Another criticism of Henning's views has been made by Pradines in a paper entitled *On the objectivity of odours and tastes.* No experimental evidence is given in this study. The author maintains that olfaction and gustation can only be understood if they are examined in relation to the satisfaction of bodily needs and that, in fact, this relation is the paramount consideration involved. Henning's initial error, he says, is excessive generalization.

Psychological factors, however, have too frequently been neglected by those who endeavour to develop theories despite the fact that much valuable information concerning olfaction has been acquired by psychologists *(e. g.,* Börnstein, Jellinek). They have studied, for example, such subjects as the effect of odours on memory, on learning, on productivity and on sales promotion.

According to Thompson, a satisfactory theory, if and when it is developed, will be based on knowledge of detailed molecular and spatial configurations of olfactory pigments and organs and of odorant molecules. Like Guillot, he rejects all vibration theories. A satisfactory hypothesis, it may also be assumed, would be in harmony with those aspects of information theory that are eventually shown to be conclusive. Von Skramlik has calculated, taking account of both qualitative and quantitative differences, that the total number of odour sensations is many millions, perhaps hundreds of millions, and Hainer and his colleagues, using such factors as the number of olfactory receptors and neurons, and the capacity of the olfactory nerves to transmit information, estimate that, if there are at least 10,000 known odours, the (human) olfactory system could distinguish 16,000,000 "patterns or odors". It was Hainer, too, who rightly emphasized that reluctance to give due weight to the indications of information theory has retarded advance in the study of olfaction and he has speculated that "a perceived odor sensation is a rapidly repeating, parallel functioning, 24-binary unit code of about 30 just noticeable differences each with greater than a 52% increase and with a dynamic range of about 200,000 times threshold concentration".

Information theory is said to suggest, also, that there are indeed, certain primary odours and that they may probably number more than twenty-five (for the better equipped species). Perhaps this suggestion is to some extent in harmony with the results of the histological study reported as long ago as 1924 by Estable. He suggested that the olfactory nerve fibres do not have identical functions and that the nerve, in fact, is equivalent to a plurality of nerves. If this is so, he said, it provides an explanation of the "pluralité d'énergies olfactives spécifiques". Possibly also, it may become possible to formulate olfaction (and other biological phenomena) in mathematical form as Elsberg did, surely prematurely, in 1939 and Narodny in 1957. "Every odour", said the latter, "can be represented by a vector in a three-dimensional space of rectangular cartesian co-ordinates".

Solubility in Air

Reference has been made in this section to the solubility of odorants in air and, obviously, some comment on this must be made. It has been suggested by Erdmann and certain other continental workers that, for a substance to be odorous, it must be "soluble" in air. Other investigators have regarded this as meaningless holding that there can be no specific "solubility" of one gas or vapour in another. Nevertheless, Erdmann's experiments appear to show that certain odorous substances at least behave towards air in a manner not shared by substances having little or no odour. Erdmann found, for example, that ether and liquid air are not miscible and that a drop of ether, allowed to fall into a quantity of liquid air, solidifies without dissolving. This behaviour contrasts strongly with that of liquid and solid carbon dioxide which are readily soluble in ether. Again, alcohol, water and carbon dioxide are insoluble, or almost insoluble, in liquid air where they immediately solidify and can be removed by filtration. In view of all this, it is the more remarkable that various odorous substances are appreciable soluble in liquid air so that, when the liquid is transformed into gas, this has quite a strong odour. Erdmann experimented with citral, ionone and rose oil. He allowed a drop of the odorous substance to fall into 10—20 c. c. of liquid air and then transferred the vacuum flask containing the mixture to a room which was free from the odour of the substance under examination. In this second room, the mixture was filtered and the clear filtrate in a vacuum flask, was transferred to a third room where persons who had not participated in work with the odorous substance tested the odours. On shaking the vacuum flask, the weak odour at first perceived became very strong and the strength increased as the nitrogen content of the liquid decreased and its oxygen content consequently increased. Erdmann proposed to continue the experiments but seems not to have done so. In any case, his experimental evidence has not been refuted and he had, no doubt, some justification for deducing, from the solubility of odorous substances in liquid air, that gaseous air likewise specifically "dissolves" these substances.

Perhaps a quite recent report by Bartlett and Phillips may appropriately be mentioned here. They tell that liquid oxygen now much used in aircraft, may contain undesirable odorous impurities some of which may be present in micro-crystalline form.

Male and Female — Young and Old

> La sensibilité olfactive brute se développe jusqu'à l'âge de six ans, puis diminue progressivement. Au contraire, la perception, la différenciation de l'eau distillée et de l'eau camphrée, la reconnaissance des odeurs différentes deviennent de plus en plus fines et sûres avec l'âge. On comprendra que, comme toutes les fonctions du système nerveux, l'olfaction se développe dans les premiers âges de la vie... La femme a une olfaction meilleure que l'homme... En outre, nous avons constaté que la femme est, sous le rapport de l'olfaction plus précoce que l'homme. *Toulouse et Vaschide*

As far as we can tell at present, no appreciable racial differences in olfactory acuity exist nor are there marked unacquired differences between normal persons and those, otherwise normal but deprived of senses other than that of smell. Disease, of course, has its effect and a considerable body of evidence on the point has been published. It might be expected, too, that sex differences and differences due to age would perhaps exist. Some, but noy much, light has been thrown on these questions. As regards sex differences, apart perhaps from those reported by Le Magnen and other French workers to which reference will be made later, the evidence is distinctly conflicting. To some extent, if not entirely, this may be due to the fact that the investigators have used too few subjects and too few odorous substances so that allowance could not be made for education, familiarity and other factors.

Male and Female

> Wie bei vielen anderen Sinnen, ist auch der Geruchssinn bei Frauen besser entwickelt als bei Männern aber minder intelligent. *Klimont*

About eighty years ago, E. H. S. Bailey and his co-workers, at a meeting of the American Association for the Advancement of Science, described experiments designed to test the delicacy of the sense of smell in men and women. The full account of these investigations, it seems, remains unpublished but two brief reports of them were printed, one in the U.S.A. and, later, one in England. The test substances mentioned in the earlier report were aqueous solution of oil of lemon 1 : 4000, aqueous solution of wintergreen oil 1 : 2700, aqueous solution of hydrocyanic acid 1 : 512 and, for some reason which is not apparent, aqueous solution of potassium hydroxide. The solutions "were successively diluted so that each bottle was one half the strength of the previous one in the series, and the dilution was continued till it was impossible to detect the substances by smell". The number of persons tested, 48 (27 male and 21 female) was regarded as too small to justify absolute conclusion but it was nevertheless clear that the males were definitely superior to the females in the delicacy of the sense of smell. Thus, while the male averages for the minima perceptible of hydrocyanic acid, lemon oil and wintergreen oil were 1 : 112,000, 1 : 280,000 and 1 : 600,000 respectively, the corresponding female averages were 1 : 18,000, 1 : 116,000 and 1 : 311,000. Bailey and Powell noted that detection and differentiation were improved by familiarity with odours and by education of the sense of smell.

The later report gives an account of tests conducted in the same way but with only thirty-four observers (17 male, 17 female). This time, the substances employed were clove oil, amyl nitrite, extract of garlic, bromine and potassium cyanide. Again, as the authors themselves say, the averages showed "that the sense of smell is in general much more delicate in the case of male than of female observers". That this conclusion is justified can be seen from the averages found. They were: clove oil, male 1 : 88218 and female 1 : 50667; amyl nitrite, male 1 : 783,870 and female 1 : 311,330; extract of garlic, male 1 : 57927 and female 1 : 43900; bromine, male 1 : 49254 and female 1 : 16244; potassium cyanide, male 1 : 109,140 and female 1 : 9002.

The persons tested by Bailey and his collaborators were presumably healthy but this is not stated nor are any ages given. Many of those examined by Ottolenghi and Lombroso were "born" criminals but they were healthy and they were non-smokers or did not smoke heavily. The odorant used was aqueous solution of essence of cloves in concentrations ranging from 1 : 100 to 1 : 50000. It seems clear from the results recorded that the criminals were much less sensitive to the odour than were others but the statements made concerning differences between men and women are confusing and, to some extent, they seem to conflict with the results given.

The American and Italian experiments were followed by those of the French investigators Toulouse and Vaschide who conducted four series of tests. Forty-four men and an equal number of women were examined in the first series. Their ages ranged from 20 to 36 and they were all inmates of an institution. One test substance only, camphor, was used (in water). For the men, its minimum perceptible was 9 parts in 100,000 whereas for the women the value was 1 : 100,000. The number of persons examined in the second series was 43 and of these 16 were women. They were allowed to smell nine odorous solutions (orangeflower water, cherry laurel water, rose water, and alcoholic solutions of citral, artifical musk, essence of mint, anethole, and camphor). Again the females were superior to the males in their power to recognize these substances.

In his third series of tests, Vaschide was concerned chiefly with the effects of old age. Accordingly it will suffice, at this stage, to say that the average age of his observers (again inmates of an institution) was 78 and that the thirty women were superior to the men. As we shall see, Toulouse and Vaschide also studied the olfactory acuity of children. In doing so, they naturally paid attention to sex differences and reached the following conclusion: "Women have better olfaction than men. This seems to us undeniable and rests on the examination of 237 individuals of all ages, individuals who have been studied from the point of view of four phenomena (sensation, perception, recognition of water, recognition of different odours). In these four series of researches, each of which represents an average of twenty individual experiments or, in all, about 20,000 data, feminine superiority has always appeared evident. We believe we have proved this decisively. Besides, we have proved that, in respect to olfaction, women are more precocious than men."

Like the American workers, Toulouse and Vaschide obtained a considerable number of anomalous results and encountered cases of partial or complete anosmia. (Vaschide states that Pasteur was completely anosmic.) With regard to the possible effects on olfactory acuity (especially in men) of smoking and drinking on which they comment, the French workers recall that amongst the infants (who neither smoked nor drank!) studied by Garbini the girls were again superior to the boys. A rather serious objection to the French work is that the persons examined may have been unsuitable for reasons of ill health.

It would be tedious to continue to give details of the reports, about thirty in number, published since the time of Toulouse and Vaschide. Some of them state that the olfactory acuity of women is superior to that of men, some state the opposite and some affirm that there are no sex differences. Henning claimed that there are sex differences but that they are simply due to differing experiences, women being superior to men in smelling food and flowers and men being superior to women in smelling chemicals and pharmaceuticals. A similar opinion was expressed by Kenneth and, much later, by Zilstorff-Pedersen. Perhaps the most trustworthy experiments in this field are those of Schneider and his co-workers who tested 46 persons (14 male, 32 female) aged 18 to 45 in an olfactorium, using citral as odorant. They concluded that, in detecting this substance, women were about twice as sensitive as were men.

It is clear that, for the present, we are far from being in a position to say whether or not one sex is superior to the other in olfactory acuity. And the very interesting findings of Le Magnen and Guillot, although not fully confirmed — a few isolated reports have adduced some supporting evidence — suggest that partial anosmia, which may greatly obscure the comparison, may be a normal physiological condition and that physiological hormone action and the changes to which it is subject, especially in women, can have a profound effect on ability to detect and to appreciate the quality and strength of the odours of certain substances, particularly sex hormones or related materials. This action, which naturally affects females much more than males, may produce very great periodic variations in sensitivity and there are also diurnal variations. Men and women, boys and girls, may differ greatly in their ability to smell these odorants and may differ greatly also in the qualities which they attribute to the odours, in some cases exhibiting partial or total anosmia. Even persons trained in the perception of odours (composers of perfumes, for example), the French investigators say, may be completely anosmic to one or two odours although they perceive other odours normally. Such unperceived odours include those of ionone (violets), hydrocyanic acid (almonds) and trinitro-*tert*-butyltoluene (musk). More surprising is the fact that persons anosmic to one "almond" or "musk" odour are not by any means necessarily anosmic to another and apparently very similar "almond" or "musk" odour. The same holds for the camphoraceous odours. Persons who cannot smell nitrobenzene may smell the very similarly smelling benzaldehyde. Again, there is at least one substance, sclareol, to which most people are anosmic. Most men, moreover, and most children of both sexes are anosmic to the odour of exaltolide (14-hydroxytetradecane carboxylic acid) whereas most women find this odour strong or even violently strong, the strength varying with the stage in the monthly cycle [1]. In both men and women, olfactory sensitivity to a particular compound can, in some cases, be very greatly increased by a single intramuscular injection of that compound and the increase may sometimes persist for a very long time (thirty days, for example). Sensitivity, too, depends or *rate* of inhalation, but is to a large extent independent of the volume of odorous air inhaled. Within certain limits, when rate of inhalation is doubled, the threshold concentration is halved. In the hormonal control of olfactory sensitivity, androgens and oestrogens exhibit antagonistic effects. In women, absence or removal of the ovaries may produce anosmia to

[1] In 1967 Vierling and Rock tested the sensitivity of 73 student nurses to the odour of exaltolide and reached conclusions similar to those of the French workers.

exaltolide or may raise the threshold a hundred or a thousand fold whilst administration of oestrogen may restore normal sensitivity.

It follows from all these interesting findings that identity of olfactory sensation may not be a criterion of identity of sensory and psycho-physiological mechanism affected. There are possibly several distinct olfactory mechanisms each concerned with a single odour quality and, similarly, there may be several types of partial anosmia corresponding to odours that may probably be fundamental. Consequently, subjective identity of odour need not indicate identity of sensory mechanism involved and, in fact, the odours of most, if not all simple substances may be complex. Thus there appear to be several fundamental musk odours, for example, and the total number of fundamental odours is possibly several hundreds. The similarity or identity of the odours of substances of entirely different chemical structure may be illusory, for these odours may be fundamentally different and affect independent receptors which may even differ visibly although this is rather unlikely. Partially anosmic people may lack the specific sensory equipment needed to perceive certain odours. Perhaps, if tracer elements are used, the topography of the olfactory mucous membrane may, some day, be mapped and the composing of perfumes may become a science rather than an art, the perfumer being able to predict as can the organic chemist. It may be found, too, that certain molecules may act on several types of olfactory receptor simultaneously, that various parts of the same odorant molecule may act successively, or even that certain molecules, when adsorbed on the membrane, may undergo deformation in such a way that they acquire power to stimulate the sensory cells. In these cells, moreover, or in some other part of the olfactory apparatus, there may occur sensitizing agents, olfactory antagonists, or both.

A simpler explanation of some of the phenomena mentioned above has been suggested by Stoll. He points out that a considerable number of the odorants involved are at or near the upper limit of molecular weight for olfactory perception, and he suggests that people vary in their power to perceive odours as this limit is approached.

Perhaps it should be added here that, before Le Magnen and Guillot, Blakeslee had found that he and others differed very greatly in their ability to perceive the odours of flowers, especially verbena flowers which, to some people, were quite odourless and that Kloek, who tested 850 subjects, later showed that steroid sex hormones and their metabolites have odours resembling those of human and animal products, (*e.g.*, civet, musk). As has already been mentioned (see p. 75), Kloek has pointed out that the odours of these hormones may play an important part in facilitating the tracking of human beings by dogs.

Young and Old

The olfactory powers of infants have seldom been studied. In 1859, Kussmaul tested the reactions of about twenty new-born children to the odours of asafoetida and bone oil. He found it impossible to obtain satisfactory results while they were awake and so tested them during sleep. Both the odours caused the infants to purse the eyelids, to "make faces", to become restless, to move head and arms, and to awake. When the odours were removed, the infants fell asleep again. But the response to the test could be elicited only a few times in succession because rapid adaptation occurred.

About fiteen years later, Genzmer conducted similar tests with about the same number of infants and reached similar conclusions. It is improbable, he affirmed, that infants recognize their mothers' breasts by smell during the first eight days of life. (Evidence that they do so at 32 days was obtained by Darwin who observed the lip and eye movements of one of his sons at this age.) Since Genzmer wrote, about a dozen other studies of olfaction in new-born and older infants have been conducted. Although some of these studies suggest that the human sense of smell is at first poorly developed, most of the results obtained show that, even immediately after birth, infants respond to the odours of such substances as anise, asafoetida, camphor, cloves, coffee, chocolate, orange, peppermint, pyridine, roses, violets and human and cow's milk and that, in some cases at least, the odours produce changes in heart rate and respiration. Infants, in fact, seem to react very much like adults, showing pleasure and displeasure and being readily fatigued. And some investigators have recently claimed that their olfactory acuity may increase four-fold during the first four days of life.

With children also, a few olfactory studies have been made. Henning's opinion that their acuity does not differ much from that of adults, seems to be generally accepted but it appears that very young children tolerate bad odours appreciably more readily than do older children and adults. Olfactory function it appears, is fully developed before the third year of life. The experiments of Toulouse and Vaschide, although published as long ago as 1899, illustrate this well. One hundred and sixty-three primary school children were divided into three groups (3—5 years, 6 years and 12 years) and their minima perceptible for camphor in water were determined. Like the adults tested by these authors, they were also asked to state if they could recognize the odours of nine other substances and to demonstrade their power to discriminate between distilled water and highly diluted aqueous solution of camphor. The following were the conclusions reached. "Indeterminate" olfactory sensitivity (perception without regocnition) increases with age until the age of six years is attained and then diminishes progressively. Power to recognize a specific odour, to differentiate between distilled water and highly diluted odorous solution, and to recognize numerous odours continues to grow more delicate and more sure as middle age approaches. At all ages, olfactory acuity is beyond doubt greater in females than in males and, as far as olfaction is concerned, females are more precocious than are males.

Having studied olfactory acuity in childhood and in adult life, Toulouse turned his attention to old age. For this purpose, his subjects were thirty-six men and thirty women and, as we have remarked earlier, their average age was 78 years and they were inmates of an institution. From the rhinological point of view, however, they were free from defects. Camphor was again the odorous material used. Two results clearly emerged. First, that acuity is greatly diminished in old age. And, second that female acuity remains superior in age as in youth. The diminution recorded in both men and women was from a few parts of camphor in 100,000 to a few parts in ten.

Most of the small remaining number of studies of the effects of age on human olfactory acuity published before or since the time of these French authors amply confirm the first of these findings. Boccuzzi's results (1962) even suggest that loss of acuity may begin as early as the 40th year of life and it is interesting to note that, according to Smith (1942), the percentage of atrophied olfactory nerves in man increases from 8 at 6—15 years to 73 at 76—91 years.

When the experiments on the effects of sex and age are being considered, it has to be borne in mind that there was no uniformity in the methods, apparatus, odorants and subjects used. Some of the methods, especially the earlier ones, were not very satisfactory. Amongst the odorants employed, there were irritants such as ammonia and acetic acid. Some of the subjects were not in good health. And the investigators used no olfactometer or more than one type of olfactometer. All these factors have naturally had a considerable and sometimes a detrimental influence on the results.

It is interesting to note, at the conclusion of this chapter, that Engen and Lipsitt, in 1964, like Kussmaul in 1859, have been testing newborn infants with asafoetida (and with anise oil, amyl acetate and heptanal) and observing their breathing responses [2]. The decrement of response on repeated testing that they have observed is attributed to habituation rather than to fatigue.

The large number (over 1000) of persons aged from ten to ninety years examined by Matzker in 1965 did not include newborn infants. However, using benzene, coffee, peppermint and another odorant, he found that olfactory acuity decreased with age, smokers at all ages over fifteen having lower acuity than nonsmokers. And, in the same year, Moncrieff, having experimented with large numbers of subjects and odorants, found that

[2] Engen (1965) has also used changes in infant's respiration as an index of the intensity of odour of the homologous alcohols (with from two to ten carbon atoms).

olfactory preferences exhibit distinct changes with age. Thus children prefer fruity odours while adults prefer the flowery types.

Some Curiosities

> "And now I will shut up this chapter with an instance, that some will think, perhaps, no less strange than any of the rest, which is that though they that are skilful in the perfuming of gloves, are wont to imbue them with but an inconsiderable quantity of odoriferous matter, yet I have by me a pair of Spanish gloves, which I had by the favour of your fair and virtuous sister that were so skilfully perfumed, that partly by her, partly by those that presented them her as a rarity, and partly by me, who have kept them several years; they have been kept about eight or nine and twenty years, if not thirty, and they are so well scented, that they may, for aught I know, continue fragrant divers years longer. Which instance, if you please to reflect upon, and consider, that such gloves cannot have been carried from one place to another, or so much as uncovered, as they must often have been, in the free air, without diffusing from themselves a fragrant atmosphere, we cannot but conclude these odorous steams to be unimaginably subtil, that could for so long a time issue out in such swarms, from a little perfumed matter lodged in the pores of a glove, and yet leave it stocked with particles of the same nature; though, especially by reason of some removes, in which I took not the gloves along with me, I forgot ever since I had them, to keep them so much as shut up in a box."
>
> *Robert Boyle*

Auf seinen Nasen schreitet
einher das Nasobēm,
von seinem Kind begleitet.
Es steht noch nicht im Brehm.

Es steht noch nicht im Meyer.
Und auch im Brockhaus nicht.
Es trat aus meiner Leyer
zum ersten Mal ans Licht.

Auf seinen Nasen schreitet
(wie schon gesagt) seitdem,
von seinem Kind begleitet,
einher das Nasobēm.

Morgenstern

Robert Boyle's gloves were not the only historical objects which were remarkable for the tenacity with which they retained their odour. There were also, for example, the documents of Albrecht von Haller (1708—1777), the renowned Swiss anatomist, botanist, physician, physiologist and poet, the residence of Sir Isaac Newton and, no doubt, not a few others. Von Haller recorded his finding in his celebrated *Elementa Physiologiae Corpori Humanis* (Lausanne: 1763) and Dyson has recently stated that a few grains of musk sufficed to perfume for several years the study of England's greatest natural philosopher. And it was in 1961, just three centuries after *The Sceptical Chymist* was published, that Greeley in the U.S.A. told of his sandalwood paper cutter, still emitting odour after one hundred years, and of his nose calculated to be capable of detecting one fifty-billionth of an ounce of the vapour of sandalwood oil.

Another XVIIIth century philosopher, albeit of a different school. was the French writer Étienne Bonnot who imagined a 'living statue' and described it in his famous *Traité des Sensations,* published in 1754. Bonnot, who was born in 1715, became Abbé de Condillac but, apparently, the duties of this position were nominal. His doctrine of ideology, sensationalism or sensualism, which is expounded in the book, was formulated in consequence of discussions which he had with his counsellor, enlightened critic and revered and talented friend, Mlle Ferrand to whom he dedicated the work. Because she had felt it necessary to examine the human senses one by one in order to distinguish precisely the ideas which we owe to each, he imagined a living statue, constructed internally like a human being but with a mind which, initially, had no ideas of any kind. "We supposed the marble exterior of the statue, "he wrote", to prevent the use of its senses, and we reserved to ourselves the right to open them at will to the different impressions of which they are susceptible." The first sense to be 'opened' was that of smell "because of all the senses, it is the one which appears to contribute least to the cognitions of the human mind"! Accordingly, the opening chapters of the treatise deal with the first cognitions, understanding, desires, passions, love, hate, hope, fear, will, ideas, sleep, dreams and personality of a man having no senses other than that of smell. Each of the other senses is then considered in the same way, first separately and then, as is done also with smell, in co-operation with the other senses taken one, two, three and four at a time until, finally, the statue is in possession of all five senses.

Condillac, the name by which Bonnot is now generally known, cannot be numbered amongst the world's greatest thinkers. But he was a savant of no mean stature and his Galatea-like statue, however curious, is not to be dismissed simply as an object of amusement or entertainment. Rather is it a valuable subject for philosophic speculation and all those who are interested in olfaction would do well to study it.

It was not only in the XVIIIth century and before that remarkable observations were made and strange things proposed and done. We recall Aronsohn's peculiar attempt, in 1894, to devise a nomenclature for odours (see p. 137) and we can still faintly perceive, after more than forty years, the perfume of the special ink with which Sansot's charming little book in the 'Petite Bibliothèque des Parfums' entitled *Essai sur les Parfums* (Paris: n. d.) was printed. We have heard of the proposed perfuming of underground stations and trains, of perfumed rain in Paris, of perfumed shampoos for dogs, and so on. Moreover, we know from the works of Monéry, Estève and others, that scientific books and periodicals are not the only literary sources of interesting and, in some cases at least, valuable information about osmics. From these sources, not a little amusement and entertainment can also be derived. The present chapter draws attention to several of them.

When Helen Keller, the distinguished blind and deaf writer and philanthropist paid her first visit to Germany (1956) she told representatives of the press in Munich how each city she visited impressed her by its smell. London, in winter, she said, made an impression of smog and fog, Paris one of a mixture of wine, tobacco, perfume and bread, Tokio one of spices and kitchen odours and, in Munich, she sensed an atmosphere of sagas, legends and music. Others have doubtless published comments on the odours of these cities and it may be remarked that some of the comments have been considered worthy of appearing in journals of the highest standing. About seventy years ago, two columns of the *British Medical Journal* were devoted to an anonymous discussion of "The Smells of London" and, in 1912, "Les Odeurs de Paris" were the subject of a communication to the French Académie des Sciences. In both cases, alas, it was stinks only that were discussed. The Londoners suffered from stenches from bad sewage systems and from organic refuse incompletely burned at brickworks whilst the Parisians were the victims of effluvia from organic material used in the manufacture of superphosphates.

Those who are not blind will find it difficult to understand how an 'atmosphere of music' can be apprehended by means of the sense of smell and perhaps Helen Keller's statement was chiefly of a poetical nature. It is well known, however, that some attempts have been made to trace a connection between odours and music. The gamut of odours of G. W. Septimus Piesse, perfumer, analytical chemist, author of various works on chemistry and, presumably, member of the former famous firm Piesse & Lubin of New Bond Street, London, was first described in *The Art of Perfumery and the Method of Obtaining the Odours of Plants. With Instructions for the Manufacture of Perfumes for the Handkerchief, Scented Powders, Odorous Vinegars, Dentifrices, Pomatums, Cosmetics, Perfumed Soap, Etc. to which is added an Appendix On Preparing Fruit-Essences Etc.* (1855) a book which was reprinted many times. But it is doubtful if the gamut ever has had, or could have had, any practical application. On at least two occasions, however, public performances have been given of music accompanied by the dissemination of odorous material. Scholes in the *Oxford Companion to Music* (5th Ed.: 1944), gives an account of these performances. The first took place in Paris in 1891 when the work performed was "The Song of Solomon, a Symphony of Spiritual Love in Eight Mystical Devices and Three Paraphrases", the 'book' being by Paul Roinard and the 'musical adaptations' by Flamen de Labrely. The colour of the stage decoration and the parfume used were, apparently, varied throughout the concert in accordance with the musical key. 'White Violet' was one of the perfumes employed. One year later, there was held in New York the 'First Experimental Perfume Concert in America'. This was a 'melody in odours' entitled "A trip to Japan in Sixteen Minutes" conveyed to the audience by a succession of odours. Entertaining experiments of this kind, it may be guessed, could readily be devised but, as far as can be seen at present, any other than fanciful connection between music and odours can scarcely be supposed to exist.

Fancy, it is true, is still to be found rather plentifully in the literature of osmics. One interesting and attractive, but probably little known example of this is P. J. Gonnet's booklet *Le Miel. Parfum-Saveur, Arôme* which was issued in 1951 at Montfavet (Vaucluse) with the sub-title *Contribution à l'Olfactique, à la Chimie des Parfums et Arômes et au Symbolisme des Nombres. M.* Gonnet, whose knowledge of his subject is obviously wide and deep, has much to tell of bees and men, of the world of the senses, of taste and savour, of smell, perfume and aroma. Moreover, as his sub-title suggests, he probes the 'super-psyche' of the hive and uncovers (or so he believes!) relationships between the 'magic' numbers three and six (2×3), the hexagonal cells of the compound insect eye and of the honey-comb, the modern electronic theory, the three dimensions, the six senses (he adds one), the last theorem of Fermat, the 'wisdom of the east', the 'six primary odours' and so on. He adopts, it would seem, a vibrational theory of olfaction similar in some respects to the theories of Heyninx and Beck and Miles. But his view is a curious mixture of science with something which is sometimes like mysticism and, not infrequently, borders on the occult. "C'est par l'arôme", he says, "que la terre se spiritualise et pénètre par la voie directe dans les régions supérieures de l' Être permettant du même coup à l'homme de se réaliser en se dépassant."

An older and rather different kind of fancy (one which might also be called musical, perhaps) was that of the German philosopher, poet and author Kurd Lasswitz (1848-1910). One of his writtings, *Bis zum Nullpunkt des Seins* (Breslau: 1878), a tale of the year 2371, tells the story of Aromasia Duftemann Ozodes, the celebrated ododistin, and her fiancé Oxygen Warm-Blasius, a distinguished weather controller. Musical performance having become so perfect that the human ear can tolerate but little of it, the ododion, an instrument invented by the Italian, Naso Odorato in the year 2094 has largely replaced musical instruments. By touching the keys of the ododion, odours of all kinds can be harmoniously released and caused to diffuse through the odoratorium. Physiologists have discovered the laws of osmics, chemists have synthesized innumerable odorous substances (including the exceptionally delicate perfume zwiblozin, discovered in the year 2369), and several great aromatists, amongst them Stinkerling, Riechmann, Frau Schnüffler and Aromasia's father, Duftemann, have brought the ododistical art to a high pitch of perfection. Aromasia, also, is a composer of the first rank. Her aroma compositions can ensnare the spirit with overwhelming power. But whilst *she* holds that the basis of all progress is art, and especially ododistics, Oxygen strongly believes that on science only can real progress be based. This sharp difference of opinion eventually leads to a violent quarrel between the two. Aromasia's friend, Magnet Reimert Oberton, the technical and scientific poet, who shares her opinion on progress, writes a poem condemning Oxygen's views and behaviour and the lady herself composes an ododistical accompaniment. Arrangements are made for her to perform the work in a great odoratorium to which, on the appointed evening, excited supporters of both sides flock in great numbers. Scarcely has the performance begun when a mysterious chemical wrecking device, projected through space by the angry Oxygen, penetrates the ododion, liberating sternutatory gases and foul odours. A stampede from the building follows, interaction of some of the gases induces spontaneous combustion and, in the devastating blaze which then destroys the structure, Aromasia is burned to ashes. The horrified Oxygen, who had intended only to estrange her from her ododion as a result of his attack with repulsive smells, bitterly repents his action. In complete despair, he applies his scientific knowledge so as to free himself from the power of gravity and vanishes into the darkness of outer space.

A different, and possibly more sober flight of fancy forms the basis of the modern novel *The New Pleasure* by John Gloag which was published in 1933. He supposes that Professor Adrian Frankby, F. R. S. has discovered *voe,* a substance which, when taken like snuff, renders the human olfactory membrane hyper-sensitive thus enabling those who use it to enjoy a new pleasure whenever their surroundings provide an attractive aroma. Perfumers and others at once realize the commercial significance of the discovery and large scale manufacture and sale of voe soon begins. But the hypersensitivity produced by the new substance is not confined to pleasant odours only and soon great numbers of people are crying out against the malodours which they encounter so often and so widely. Bad sanitary conditions and high degrees of atmospheric pollution, for long readily tolerated, quickly become anathema to the multitudes. An immense exodus of people from 'dirty' to 'clean' trades and from town and city to the country begins. The gigantic social upheaval which is thus caused leads to serious conflicts, especially since very powerful vested interests are inevitably threatened and since, indeed, many industries are actually completely ruined. The 'revolution' which voe causes in England spreads to other parts of the world. In the U.S.A. the damage done to the oil and automobile industries (producers of bad odours and fumes!) and a new outbreak of anti-negro agitation almost cause a second civil war. The repulsive fertilization methods of China and the Far East generally are gradually abandoned and very great hygienic progress is made everyhere. Eventually, many of the threatened industries mend their olfactory ways and are restored to a flourishing condition, all opposition to voe is withdrawn, and Frankby, for long ignored or attacked, receives belated recognition. But before this happens, the most extraordinary, unexpected and far-reaching social and economic changes due to the use of voe occur throughout the world and even in the fields of architecture, town and country planning, morals and religion, for example, the effects of the new pleasure are felt. Long after his death, the discoverer is canonized (Saint Adrian) as one of the great benefactors of the human race. He made the world a place fit to live in!

Another exercise of imagination, again quite different from that of the French, German and British writers, was that of the great Russian novelist Nikolai Vasilievich Gogol (1809—1852) when he wrote his short story *The Nose*. This bizarre narrative tells how Ivan Yakovlevich, the barber, finds in the newly baked loaf on which he is breakfasting, a human nose! This nose, is transpires, belongs to the collegiate assessor Kovalyov whose distress and horror at the loss of such a valuable and decorative organ are naturally profound. For very different reasons, the barber also suffers great distress and fear which are not dispelled when he succeeds, unobserved, in throwing the nose from a bridge into the river Neva. But after a number of unsuccessful attempts (which provide amusement for the reader), the assessor recovers his nose even if he does not necessarily 'live happily ever after'.

Dr. Vaughan-Trotter, the story of whose rise and fall constitutes another curiosity, certainly lived anything but happily after his decline. An account of his (presumably fictitious) life was given in 1898 by P. Lafargue. Although "a pretentious dolt, an inveterate discoverer of mares' nests who made impossible blunders in the simplest cases", he frequently scored astonishing hits in diagnosis, outshining the most distinguished physicians of his day. He rose with meteoric celerity to the height of fashion as a West-end practitioner in consequence of the patronage of a nobleman whose reputation he shielded at a time when terrible disgrace threatened. Vaughan-Trotter, when attending the hysterical daughter of the nobleman, discovered that she had committed murder, probably during an attack of epileptic mania. But he kept the secret. His own secret was not, as many believed, his intuition, but his nose and it was by perceiving an odour on the sick girl's glove that he made his discovery. For he had been born "with a sense of smell so acute and imperative that it quite overshadowed all his other sense perceptions". Thus he could name an animal by the smell of its blood and, professionally, could often detect in a moment what others could only uncover if at all, by careful reasoning. But it was this very power which led to his downfall. His olfactory gift became a curse. Every gutter seemed to send a poisoned arrow into his brain. Life, for him, became one long, hideous, olfactory torture. Attempts to deaden his cruel sense were ineffective. He had to abandon his profession and emigrate, a broken man, to a lonely part of Australia in the hope that he might there be able to control his environment and so to evade the lunacy or suicide which seemed imminent.

Our last 'olfactory curosity' differs completely from all the others. It was published under the title *Arithmetic by Smell* in the *"Psychological Review" (1,* 61—62, 1894) by Sir Francis Galton, F. R. S. (1822—1911), the famous founder of the science of eugenics. Taking brass cylinders having a nozzle at one end and tightly joined to a closed rubber tube at the other, he inserted loosely into the cylinders pads of cotton wool on to which odorous material (pepper-mint, camphor, phenol, aniseed, ammonia and some others) was profusely poured. With his eyes shut, he operated this simple apparatus by making a sudden grasp on the rubber tube and, at the same time, suddenly sniffing the air expelled from the nozzle. In this way, he says, "I taught myself to associate two whiffs of peppermint with one whiff of camphor, three of peppermint with one of carbolic acid, and so on. Next I practiced at some small sums in addition; at first with the scents themselves, and afterwards altogether with the imagination of them. There was not the slightest difficulty in banishing all visual and auditory images from the mind, leaving nothing in the consciousness besides real or imaginary scents. In this way, without, it is true, becoming very apt at the process, I convinced myself of the possibility of doing sums in simple addition with considerable speed and accuracy solely by means of imaginary scents. Further than this I did not go, so far as addition was concerned. It seemed a serious waste of time to continue the experiments further, because their difficulty and complexity rapidly increased". He did not attempt multiplication by smell but found that subtraction succeeded as well as addition. In a few other experiments with taste using salt, sugar, citric acid and quinine, he convinced himself that arithmetic by taste was as feasible as arithmetic by smell.

Before leaving the subject of curiosities, we may revert for a moment to the odours of Paris and then add three other remarks. In 1895, when these odours were the subject of complaint, Gerardin decided to measure their intensity by aspirating the polluted air into a vessel, steaming for fifteen minutes, treating the condensate with permanganate and determining the proportion of unconsumed permanganate with oxalic acid. In this rather unspecific way he examined the air in the Gare du Nord, in a sausage factory and in other places.

The first remark concerns a modern (1960) New York show nameed "smell-o-vision". In this, during the exhibition of a film about Spain, odours of rose, peach, wood shavings, new bread, banana, shoe polish, tobacco smoke, lemon, oil paint, wine and so on were dispersed in the cinema to harmonize with the scenes. It is said that the experiment was a success.

Second, here is the title of an amusing paper by Genot: *"Non mais, Docteur, quelle odeur!"* This tells of a rich and distinguished man who found the mild and not unpleasant odours of a factory sickening while completely oblivious of the fact that the mansion in which he lived smelled very strongly of moth balls and furniture polish.

Finally, the title only (with a minimum of comment) of an amusing booklet by Stümpke may be mentioned. This is *Bau und Leben der Rhinogradentia* (1962) which deals with various fabulous creatures having most peculiar noses and said to inhabit a group of south sea islands. These islands, of course, no longer exist because they were submerged in a great natural catastrophe.

Concluding Note

It may possibly be thought strange that more has not been said in this book about the modern, very powerful techniques — various forms of chromatography and spectrometry, for example — now being increasingly applied, together with extremely useful methods of statistical analysis, by a growing number of investigators to the study of olfaction. The value of all these procedures should certainly not be underestimated. But, as Moulton has pointed out in his most valuable contributions to the subject, the sense of smell is much more sensitive than is the flame ionization detector of a gas-liquid chromatograph. We still cannot rely, he remarks, on micro-analytical procedures to detect all odorants present in concentrations detectable by the nose; we cannot be sure that electro-physiological processes may not be misleading; and we know no enzyme system that can respond to the very low concentrations of some odorants that are sufficient to excite the receptor. Above all, he says, we must never forget that "present theories of odor qualitiy have a major weakness. Because many of the properties of the molecule are interrelated, it is not sufficient to show that a high correlation exists between a given set of molecular properties and subjectively-derived estimates of odor quality. It must also be shown that a similar correlation does not exist with other properties. At the moment few, if any, of the odor theories seem detailled enough to indicate how such a critical test could be made."

Epilogue

"Quand je commençai de tracer ces lignes, je méditais une grande oeuvre d'art et de science à la fois. J'eusse montré le culte des parfums chez les peuples les plus reculés, leur histoire et les milles secrets dont dépend leur existence. C'était trop d'ambition et j'ai dû me borner.

Personne peut-être ne s'en plaindra. Du moins, je serai satisfait si je peux, à qui me lira, communiquer un peu de mon penchant pour la connaissance des odeurs. Cette connaissance est, à l'heure actuelle encore, si misérable et si pauvre que l'homme n'en saurait tirer la moindre fierté. Ce champ en jachère a séduit ma curiosité. J'ai soulevé quelques mottes de terre dure, défriché quelques arpents. Que d'autres, plus vaillants, plus subtils et savants aussi, m'imitent: O la belle moisson qu'ils verront se dresser à la lumière des jours à venir.

Régismanset

Bibliographical Notes

Those who wish to study osmics more fully will consult the literature mentioned in the text and the author's papers on the literature of olfaction published in *Soap, Perfumery and Cosmetics* (**24**, 782—783 [1951]) and in *Parfumerie, Cosmétique, Savons*, **10**, 202—205 (1967). They should also make use of the following (and other) sources:

Biological Abstracts
Bulletin Scientifique et Industriel de la Maison Roure-Bertrand fils (Grasse) and its successor, *Recherches*
Chemical Abstracts
Chemisches Zentralblatt
Index-Catalogue of the Library of the Surgeon-General's Office, U. S. Army
Index Medicus
Nature
Psychological Abstracts
Schimmel-Berichte (now *Miltitzer-Berichte)*
Science

and also the more important cosmetic, perfumery and psychological journals. In addition, they should utilize the following bibliographies:

"AIRKEM": Odors and the sense of smell, 320 B. C. — 1947. New York: Airkem 1952. This is probably the largest and most comprehensive bibliography of osmics ever produced. It contains about five thousand references. It was proposed to issue supplements but this, apparently has not been done. Unavoidably, the book contains some errors and inconsistencies.

BAWDEN, H. H.: A bibliography of the literature on the organ and sense of smell. *J. comp. Neurol.* **11**, i—xl (1901). Contains 885 references, including many to popular articles.

FORTUNATO, V., and P. NICCOLINI: There are about 1500 references in their bibliography (see p. 9).

HENNEBERT, P. E.: His bibliography contains about 420 references. L'Olfaction. *Acta otorhino-laryng. belg.* **7**, 103—225 (1953).

HEYNINX, A.: The number of references in his book is about 1200 (see p. 8).

KENNETH, J. H.: Osmics 1922 and 1924. A thousand references.

LARGUIER DES BANCELS, L.: In his book, *Le goût et l'odorat* (1912) he gives one hundred and seven references arranged chronologically. The period covered is 1821—1910.

McCORD, C. P., and W. N. WITHERIDGE: These authors claim that their bibliography was "the first substantially complete bibliography in the English language". This is not so. (See p. 9.) About 3100 references.

MICHELS, K. M., D. S. PHILLIPS, R. H. WRIGHT, and J. PUSTEK: The bibliography published by these authors is entitled "Odors and olfaction: a bibliography 1948—1960. It can be regarded as a supplement to the "Airkem" bibliography. *Percept. Motor Skills* **15**, 475—529 (1962). About 1400 references.

OTTOSON, D.: In his paper on "Some aspects of the function of the olfactory system" Ottoson gives a list of 230 references. *Pharmacol. Rev.* **15**, 1—42 (1963).

Saveliev, N. A.: There are 240 references in his bibliography. *Trud. Fiziol. Lab. Imp. Moskov. Univ.* 4, 121—378 (1893). (See p. 7.)

Despite the existence of these valuable and indeed essential aids, the task of the student endeavouring to cover his subject even moderately well remains far from easy. The abstract sources, although indispensible, are still too slow and too incomplete and valuable work is not infrequently overlooked or discussed very late.

Again, the scattered nature of the literature and the variety of subjects *(e. g.*, anthropology, chemistry, physiology, psychology, zoology) involved constitute formidable obstacles which can probably never be fully overcome. Then there are the language difficulties and the inaccessibility of certain parts of the literature. Thus, for example, there is undoubtedly a considerable volume of Japanese and Russian literature which can be obtained by those outside the countries concerned only with great difficulty or not at all.

It should perhaps be said here that academic workers and "pure" scientists tend to ignore technical journals whilst perfurmers and perfumery chemists, in turn, tend to ignore the journals of pure science and of the learned societies. It is sufficient to add that this is bad for the progress of science.

Bibliography

Achelis, J. D.: Geruchsstudien. Arch. ges. Psychol. **71**, 273—338 (1929).

Adachi, B.: Geruch der Europäer. Globus **83**, 14—15 (1903).

Adrian, E. D.: The sense of smell. Adv. Sci. **4**, 287—292 (1948).

— Sensory discrimination: with some recent evidence from the olfactory organ. Brit. med. Bull. **6**, 330—333 (1950).

— Olfactory discrimination. Ann. Psychol. **50**, 107—113 (1951).

— The discrimination of odours by the nose. Schweiz. med. Wschr. **82**, 973—974 (1952).

— Sensory messages and sensation; the response of the olfactory organ to different smells. Acta physiol. scand. **29**, 5—14 (1953).

— The mechanism of olfactory stimulation in the mammal. Adv. sci. **9**, 417—420 (1953).

— Il meccanismo sensoriale dell'organo olfattorio. Boll. Soc. ital. Biol. sper. **30**, 411—417 (1954).

— Des réactions électriques du système olfactif. Actualités neurophysiol. **1**, 11—17 (1959).

—, and C. Ludwig: Nervous discharges from the olfactory organs of fish. J. Physiol. **94**, 441—460 (1938).

Agricola: Met welke zintuigen van het veerwild moet de jager het meest rekenen. Nederl. Jager **38**, 568—569 (1933).

Aitken, J.: Evaporation of musk and other odorous substances. Proc. Roy. Soc. Edinburgh **25**, 894—902 (1905/1906).

Alexander, J.: Odor detection. Chem. Eng. News **27**, 2227 (1949).

— The catalyst theory of olfaction. Proc. Sci. Sect. Toilet Goods Assoc. No. **16**, 27—34 (1951).

— Catalysis and olfaction. Science **114**, 674—675 (1951).

Allen, C. G. B.: Physiological aesthetics. London: King 1877.

— The dog's universe. Gentlemen's Mag. **249**, 287—301 (1880).

— Sight and smell in vertebrates. Mind. **6**, 453—471 (1881).

Allen, W. F.: A comparative study of the olfactory and trigeminal reflexes elicited by various vapors in different mammals. J. Wash. Acad. Sci. **26**, 466 to 473 (1936).

— Olfactory and trigeminal conditioned reflexes in dogs. Amer. J. Physiol. **118**, 532—540 (1937).
ALLISON, A. C.: The morphology of the olfactory system in the vertebrates. Biol. Rev. Cambridge Philos. Soc. **28**, 195—244 (1953).
— Cyanide-smelling deficiency among Africans. Man **53**, 176—177 (1953).
AMIROV, R. Z.: Vlijanie podporogvych zapachovych razdrazitelej na cuvstviteljnostj English, 562—565.
— Cuvstviteljnostj obonjanija v pokoe i vlijanie zvukovjch razdrazitelej. Bjull. eksp. biol. Med. **48**, No. 11, 18—21 (1959); English 1326—1329.
AMOORE, J. E.: The stereochemical specificities of human olfactory receptors Perf. Ess. Oil Rec. **43**, 321—323; 330 (1952).
— The stereochemical theory of olfaction. I. Identification of the seven primary odours. II. Elucidation of the stereochemical properties of the olfactory receptor sites. Proc. Sci. Sect. Toilet Goods Assoc. No. 37, Suppl., 1—12; 13—23 (1962).
— Stereochemical theory of olfaction. Nature **198**, 271—272 (1963).
— Stereochemical theory of olfaction. Nature **199**, 912—913 (1963).
— Current status of the steric theory of odor. Ann. N. Y. Acad. Sci. **116**, 457—476 (1964).
— Specific anosmia: a clue to the olfactory code. Nature **214**, 1095—1098 (1967).
— J. W. JOHNSTON, and M. RUBIN: The stereochemical theory of odor. Sci. American **210**, 42—49 (1964).
—, G. PALMIERI, and E. WANKE: Molecular shape and odour: pattern analysis by PAPA. Nature **216**, 1084—1087 (1967).
— Psychophysics of odor. Cold Spr. Harb. Symp. quant. Biol. **30**, 623—635 (1965), discussion, 636—637.
—, and D. VENSTROM: Sensory analysis of odor qualities in terms of the stereochemical theory. J. Food. Sci. **31**, 118—128 (1966).
ANDRES, P.: Beitrag zur Frage: Kann Zucker riechen? Z. Zuckerindust. **78**, 319—320 (1953).
ANDRIEU, R.: L'Olfaction en microphysiologie humaine. France Parf. **9**, 161—166 (1966).
ANGELERI, F., and M. C. CARRERAS: Problemi di fisiologia dell'olfatto. I. II. Riv. Neurobiol. **2**, 255—274, 275—297 (1956).
ANGLERAUD, O., et F. BORELLI: Études pratiques sur l'odorisation. Intensités d'odeur des différents gaz. C. R. Congr. Indust. Gaz, **82**, 862—898 (1965). Discussion, 899—904.
ANTONELLI, A. R.: Elektrophysiologische Untersuchungen am Olfaktionssystem der Katze während intrahämatischer Injektion von Essenzlösungen. Laryngol. Rhinol. Otol. **41**, 822—828 (1962).
ARNOLD, E.: East and West. London: Longmans, Green 1896.
ARONSOHN, E.: Beiträge zur Physiologie des Geruchs. Arch. Physiol. 163—167 (1884).
— Experimentelle Untersuchungen zur Physiologie des Geruchs. Arch. Physiol. 321—357 (1886).
— Versuch einer Nomenclatur der Geruchsqualitäten. Arch. Laryngol. Rhinol. **2**, 42—47 (1894/95).
ARUNDEL, R.: The police dog: his selection and training. Police J. **1**, 423—431 (1928).
— Police dogs and their training. 4th Ed. London: Police Rev. Publ. Co. 1948.
ASAKA, H.: The studies on the objective olfactory test by galvanic skin response. J. Otorhinolaryngol. Soc. Japan **68**, 100—112 (1965).
ASHTON, E. H., J. T. EAYRS, and D. G. MOULTON: Olfactory acuity in the dog. Nature **179**, 1069—1070 (1957).

Audubon, J. J.: Account of the habits of the turkey buzzard *(Vultur aura)* particularly with the view of exploding the opinion generally entertained of its extraordinary power of smelling. Edinburgh New Philos. J. **2**, 172—184 (1827).

— Ornithological biography, or an account of the habits of the birds of the United States of America. Edinburgh: Black 1831—1839.

Auld, S. J. M.: Psycho-galvanic measurement of smell. J. Inst. Petrol. Technol. **9**, 389—391 (1923).

Austerweil, G., et G. Cochin: Sur quelques relations entre la constitution moléculaire et l'odeur. C. R. Acad. Sci. (Paris) **150**, 1693—1695 (1910).

— Sur certaines causes des odeurs géraniques. C. R. Acad. Sci. **151**, 440 (1910).

Ayrton, W. E.: Physics of smell. Rep. Brit. Ass. adv. Sci., 68th Meeting, 771—777 (1898). *Also in* Nature **58**, 450—452 (1898).

Azaretti, E.: Una classificazione fondamentale degli odori. Profum. Ital. **3**, 144—149 (1925).

Bachem, A.: The modified Fechner-Weber law as a complex law of dose-action. Amer. J. Psychol. **69**, 588—603 (1956).

Bachman, J.: Remarks in defence of [Mr Audubon] the author of the "[Biography of the] Birds of America". Mag. Nat. Hist. J. Zool. Bot. Mineral. Geol. Meteorol. **7**, 164—175 (1834).

Backman, E. L.: Experimentella undersökningar öfver luktsinnets fysiologi. Upsala Läk. Fören. Förh., N. F. **22**, 317—470 (1916/17).

— Über die Verstaubungselektrizität der Riechstoffe. Arch. ges. Physiol. **168**, 351—371 (1917).

— Några trötthetsfenomen inom lukt- och smaksinnenas omrade. Hygiea **79**, 886—896 (1917).

— The olfactology of the methylbenzol series. Proc. K. Akad. Wetensch. (Amsterdam) **19**, 943—956 (1917).

— Note sur la puissance des odeurs et leur solubilité dans l'eau et dans l'huile. J. Physiol. Path. gén. **17**, 1—4 (1917/18).

Baglioni, S.: Contributions expérimentales à la physiologie du sens olfactif et du sens tactile des animaux marins ("Octopus" et quelques poissons). Arch. ital. Biol. **52**, 225—230 (1909).

Bailey, E. H. S., and E. L. Nichols: Preliminary notes on the delicacy of the special senses. N. Y. med. J. **40**, 325 (1884).

—, and L. M. Powell: Some special tests in regard to the delicacy of the sense of smell. Trans. Kansas Acad. Sci. **9**, 100—101 (1885).

Bain, A.: The senses and the intellect. 2nd Edition. London: Longmans 1864.

Bajandurov, B. I., and E. F. Larin: K. fiziologii obonjatel'nogo analizatora ptits. Tr. Tomsk. Gosudarstven. Med. Inst. **2**, 204—224 (1935).

Baker, A. F.: The significance of smell: some speculations on the phenomenology of olfaction. Sci. J. roy. Coll. Sci. **5**, 84—101 (1935).

Baker, R. A.: Threshold odors of organic chemicals. J. Amer. Water Works Ass. **55**, 913—916 (1963).

Baldus, C.: Untersuchung über Geruchsschwellen. Dissertation. Würzburg 1936.

Bang, B. G.: Anatomical evidence for olfactory function in some species of birds. Nature **188**, 547—549 (1960).

— Localization of gustatory and olfactory enzymes in the rabbit, and the problems of taste and smell. Nature **168**, 977—979 (1951).

Baradi, A. F., and G. H. Bourne: Gustatory enzymes. Brit. J. Nutrit. **5**, vii—viii (1951).

Barail, L. C.: Cosmetic odor measurement. Amer. Perf. Ess. Oil Rev. **60**, 347—349 (1952).

Barrows, W. B.: The sense of smell in *Cathartes aura*. Auk N. S. **4**, 172 (1887).

Bartalena, G.: Olfazione mediata in anosmici; esperimenti tendenti a dimostrare il valore delle cavita nasali come laboratorio biochimico destinato all'olfatto. Boll. Mal. Orecch. **71**, 164—178 (1953).

— Olfattometria e rinomanometria. (Loro rapporti sotto l'influsso di farmaci attivi sulla permeabilita dei tessuti e sul calibro vasale.) Boll. Mal. Orecch. **73**, 31—42 (1955).

—, e G. Romeo: Olfattometria ematogena elettroencefalografica in soggetti normali. Boll. Mol. Orecch. **80**, 14—23 (1962).

— — Sulla fisiogenesi dell'olfazione ematogena. Studio elettroencefalografico. Boll. Mal. Orecch. **80**, 24—31 (1962).

Bartlett, R. G., and N. E. Phillips: Olfactory identification of liquid oxygen contaminants. Aerospace Med. **32**, 621—629 (1961).

Bartsch, P.: Birds and smell. Auk **60**, 271—272 (1943).

Bassiri, T.: Odorous and odourless. Soap, Perfumery and Cosmetics **36**, 877—879 (1963).

Bateson, W.: The sense-organs and perceptions of fishes; with remarks on the supply of bait. J. Marine Biol. Ass. U. K., N. S. **1**, 225—256 (1889/90).

Bazzoni, C. B.: Loss of weight of musk in a current of dry air. J. Franklin Inst. **180**, 463—469 (1915).

Beach, F. A., and R. W. Gilmore: Response of male dogs to urine from females in heat. J. Mammalol. **30**, 391—392 (1949).

Bear, I. B., and R. G. Thomas: Nature of argillaceous odour. Nature **201**, 993—995 (1964).

Beck, H. H.: The occult senses in birds. Auk **37**, 55—59 (1920).

Beck, L. H.: A quantitative theory of the olfactory threshold based upon the amount of the sense cell covered by an adsorbed film. Ann. N. Y. Acad. Sci. **116**, 448—456 (1964).

—, and W. R. Miles: Some theoretical and experimental relationships between infrared absorption and olfaction. Science **106**, 511 (1947).

Becker, R. F., J. E. Markee, and J. E. King: Studies on olfactory acuity in dogs. I. Discriminatory behaviour in problem box situations. Brit. J. Anim. Behav. **5**, 94—103 (1957).

—, J. E. King, and J. E. Markee: Studies on olfactory discrimination in dogs. II. Discriminatory behavior in a free environment. J. comp. Physiol. Psychol. **55**, 773—780 (1962).

Bedichek, R.: The sense of smell. New York: Doubleday & Co. 1960.

Bedini, S. A.: Holy smoke: the oriental fire clocks. New Scientist **21**, 537—539 (1964).

Bednár, M., u. O. Langfelder: Über das „intravenöse" (hämatogene) Riechen. Mschr. Ohrenheilk. **63**, 1292 (1929).

Beebe, C. W.: New World vultures. Zool. Soc. Bull. **2**, 458—461; 465—470 (1906/13).

Beets, M. G. J.: Errors in odor theory. Amer. Perfumer **55**, 290 (1950).

— Structure and odour. Soc. Chem. Indust., Monograph No. 1, 54—90. London, 1957.

— Quelques aspects du probleme de l'odeur. Parf. Cosmét. Savons **5**, 167—185 (1962).

— A molecular approach to olfaction. Molec. Pharmacol. **II** (1964).

Beidler, L. M.: Chemical excitation of taste and odor receptors. Ad. Chem. Ser. **56**, 1—28 (1966).

Békésy, G. v.: Olfactory analogue to directional hearing. J. appl. Physiol. **19**, 369—373 (1964).

BELLEVILLE, R.: Neue Versuche auf dem Gebiet der Fährtenreinheit und des Identifizierens von Gegenständen auf der Fährte. Z. Hundeforsch. N. F. **13**, 17—28 (1938).

BENHAM, W. B.: The olfactory sense in Apteryx. Nature **74**, 222—223 (1906).

BERTHELOT, M. P. E.: Sur les limites de sensibilité des odeurs et des émanations. Ann. Chim. [viii] **3**, 293—295 (1904).

BEST, M. G. S., and M. D. HAVILAND: The sense of smell in the grey lag-goose. Brit. Birds **7**, 34—37 (1913/14).

BIENFANG, R.: The subtle sense. Norman, Oklahoma: Univ. Oklahoma Press 1946.

BIERENS DE HAAN, J. A.: Sieben Jahre tierpsychologische Arbeit in Amsterdam. Z. angew. Psychol. **27**, 236; 241—242 (1926).

— Wat Gaat er om in uw Hond? Amsterdam: Strengholt 1947.

BIJTEL, J., and C. J. A. VAN ITERSON: Psycho-galvanic reflex phenomenon in sense organs, especially the nose. Acta oto-laryngol. **7**, 31—40 (1924/25).

— — Recherches sur le réflexe psychogalvanique, en particulier comme réaction aux excitations olfactives. Arch. néerl. Physiol. **10**, 447—448 (1925/26).

BINET, A., et J. PASSY: Contribution à l'étude de l'olfaction chez le chien. C. R. 24me Session, Assoc. Franç. Avance. Sci. **ii**, 659—661 (1895/96).

BLAIZOT, P.: Parfums et Parfumeurs. Paris: Editions à l'Étoile 1946.

BLAKESLEE, A. F.: Unlike reaction of different individuals to fragrance in verbena flowers. Science **48**, 298—299 (1918).

— Demonstration of differences between people in the sense of smell. Sci. Mthly **41**, 72—84 (1935).

BLINOVA, E. A.: O normirovanii koncentracii vescestv s siljnym zapachom v proizvodstvennych pomescenii. Gig. Sanit. **30**, No. 1, 18—22 (1965).

BLONDAL, H.: The sense of smell. Ztschr. Psychol. **161**, 292—295 (1957).

BLUNK, R.: Die Ausbildung des Hundes zur Spurenreinheit. Rostock: Hinstorffs Hofbuchdruckerei 1926.

BOCCUZZI, V.: Ricerche sulla soglia olfattiva nei ciechi. Boll. Mal. Orecch. **80**, 445—467 (1962).

— Fenomeni di senescenza nell'olfatto. Boll. Mal. Orecch. **80**, 654—665 (1962).

BÖRNSTEIN, W.: Über den Geruchsinn. I. Zur Psychologie des Geruchsinns. II. Zur Methodik der Geruchsprüfung unter besonderer Berücksichtigung der klinischen Bedürfnisse. III. Fälle und ihre Deutung. IV. Über die biologische und klinische Bedeutung des Geruchsinns. Über die Beziehung zwischen Symptom und Herd im Riechsystem. Dtsch. Z. Nervenheilk. **104**, 55—77; 78—91; 173—195; 195 —207 (1927).

BÖTTGER, P.: Anweisung zum Abrichten von Suchhunden auf Grund praktischer Erfahrungen. IV. Aufl. Augsburg: Verein für deutsche Schäferhunde (SV), Sitz München (EV) 1926.

— Vom Werdegang des Polizeifährtenhundes. Z. Hundeforsch. **2**, 196—222 (1932).

— Hund im Dienste der Kriminalpolizei. Krim. Monatsh. **10**, 121—127 (1936).

— Hunde im Dienste der Kriminalpolizei unter besonderer Berücksichtigung des Mordfalles Seefeldt. Z. Hundeforsch. N. F. **5**, 32 pp. (1937).

BONNOT, E. (ABBÉ DE CONDILLAC): Traité des Senations (à Madame la Comtesse de Vasse). London 1754.

BORDIER, H., et T. NOGIER: Recherches sur la cause de l'odeur prise par l'air soumis aux radiations ultra-violettes émises par la lampe à vapeur de mercure. C. R. Acad. Sci. (Paris) **147**, 354—355 (1908).

BORSANYI, S. J., and C. L. BLANCHARD: Psychogalvanic skin response olfactometry. Ann. Otol. Rhinol. Laryngol. **71**, 213—221 (1962).

BORSANYI, S. J., C. R. BLANCHARD, and F. J. BAKER: Psychogalvanic skin response as a method of objective olfactometry. Bull. Sch. Med. Univ. Maryland **47** 10—11; 13 (1962).

BOSSERT, W. H., and E. O. WILSON: The analysis of olfactory communication among animalis. J. theor. Biol. **5**, 443—469 (1963).

BOUCHER, R.: Une science du passé: l'osphrésiologie. Union Méd. Canada **76**, 1450—1453 (1947).

BOUDOUARD, O.: Les odeurs de Paris. C. R. Acad. Sci. (Paris) **154**, 238—240 (1912).

DE BOUILLÉ, R.: De l'odorat chez les vautours. Bull. Soc. Ramond, [iii], 140—144 (1875).

BOURNOT, K.: Über den Geruch und die Duftstoffe. Dtsch. Parf.-Ztg. **25**, 241 bis 243 (1939).

— Geruchsprobleme. Z. ges. Naturwiss. **5**, 196—199 (1939).

BOYLE, R.: Works, New Edition for Rivington *et al.* London 1772.

BRANDT, B.: Über den Geruch brechenden Gesteins. Naturwiss. **8**, 689—690 (1920).

BRAUN, J. v., u. H. KRÖPER: Geruch und Konstitution. I. Ber. dtsch. chem. Ges. **62B**, 2880—2885 (1929).

BRIGGS, M. H., and R. B. DUNCAN: Pigment and olfactory mechanism. Nature **195**, 1313—1314 (1962).

BRINKLEY, F.: Japan, its history, arts and literature. Boston and Tôkyô: J. B. Millet 1901.

BRODAL, A.: The present status of our knowledge of the olfactory system of the brain. Schweiz. med. Wschr. **77**, 971—977 (1947).

BRONSTEIN, A. I.: Vkus i Obonjanie. Moskva: Acad. Nauk SSSR 1950. — Lokalizacija degidraz v obonjateljnych kletkach pozvonocnych zivotnych. Dokl. Akad. Nauk SSSR **142**, 936—939 (1962).

BROUGH, E.: The bloodhound and its use in tracking criminals. London: "Illustrated Kennel News" n. d.

BROWN, B. F.: The characteristic odor of nodular iron fractures and carbide slags. Foundry **79**, No. 10, 174 (1951).

BROWN, K. S., and R. R. ROBINETTE: No simple pattern of inheritance in ability to smell solutions of cyanide. Nature **215**, 406—408 (1967).

BUDGETT, H. M.: Hunting by scent. London: Eyre & Spottiswoode 1933.

— Solving the mystery of scent. Field **163**, 170 (1934).

BÜHLER, K.: Der Atemfaktor in tierischen Geruchsspuren. Jb. Psychol. Psychother. **1**, 479—483 (1953).

BULL, H. O.: On the nature of purposive movement in fishes. Rep. Dove Marine Lab. **18**, 39—46 (1928/29).

— Studies on conditioned responses in fishes. II. J. Marine Biol. Assoc. U. K. **16**, N. S. 615—637 (1929/30).

BUNGENBERG DE JONG, H. G., u. G. P. SAUBERT: Kolloidmodelle biologischer Vorgänge. Für Dämpfe organischer Nichtelektrolyte empfindliche Koazervate als Modelle der Geruchssinnreizung. Protoplasma **28**, 329—344 (1937).

— — Models for the stimulation of the organ of smell. Proc. K. Akad. Wetensch., Amsterdam **40**, 302—306 (1937).

BURTON, R. W.: Scent. J. Bombay Nat. Soc., **49**, 116—117 (1950).

BUYTENDIJK, F. J. J.: L'odorat du chien. Arch. neerl. Physiol. **5**, 434—457 (1921).

— De Psychologie van den Hond. Amsterdam: Cosmos 1932.

— The Mind of the dog. London: Allen & Unwin 1935. English translation of *De Psychologie van den Hond.*

CALVIN, A.: Olfactory discrimination. Science **131**, 1263—1265 (1960).

CALVIN, A. D., C. M. WILLIAMS, and N. WESTMORELAND: Olfactory sensitivity in the domestic pigeon. Amer. J. Physiol. **188**, 255—256 (1957).
CAMPBELL, B.: Quantum detection in biological systems with particular reference to olfaction. Fed. Proc. **15**, 29—30 (1956).
CAMPION, D.: The perfect team. London: Robert Hale 1959.
CASTEX, M. R., y F. VIERHELLER: Estudio fisico del olor. Bol. Acad. Nac. Med. Buenos Aires 722—763 (1934); 284—317, 655—670 (1935); 825—860, 1105—1123 (1936); 31—70 (1938); 138—146 (1940); 399—421 (1941).
CATON, J. D.: The senses of sight and smell of the wild turkey and the common deer. Amer. Naturalist **3**, 28—30 (1869/70).
CHAMULEAU, F. J., et A. K. NOYONS: La pression de vapeur et le spectre d'absorption de quelques matières odorantes. Acta Brev. Neerland. Physiol. Pharmacol. Microbiol. E. A. **2**, 94—95 (1932).
CHAPMAN, A.: The sense of smell in birds. Field **118**, 1447 (1911); **119**, 48 (1912).
CHAPMAN, F. M.: Life in an Air Castle. Nature studies in the tropics. New York and London: Appleton-Century 1938.
CHARLES, W. F.: The sense of smell. Engl. Mechanics **121**, 364 (1925).
CHEESMAN, G. H.: Odour and chemical constitution — a new approach. Roy. Austral. Chem. Inst. Proc. 70—73 (1960).
CHERUBINO, M., e B. SALIS: Valutazione dell'acuità olfattiva nel cieco e nel sordomuto; ulteriore contributo al sinergismo funzionale intersensoriale. Boll. Mal. Orecch. **75**, 239—251 (1957).
CLARK, W. E. LE GROS: Inquiries into the anatomical basis of olfactory discrimination. Proc. roy. Soc. **146B**, 299—319 (1957).
CLOQUET, H.: Osphrésiologie, ou Traité des Odeurs, du sens et des organes de l'olfaction. Paris: 2 Éd. Méquignon-Marvis 1821. German translation, Weimar: Verlag des Landes-Industrie-Comptoirs 1824.
COHEN, N. E.: The relativity of absolute judgements. Amer. J. Psychol. **49**, 93—100 (1937).
COLENSO, W.: Some reminiscences of the Maoris. Nature **47**, 41—45 (1892/93).
COLVIN, S.: Fine Arts. Ency. Brit. **10**, 355—375 (1910/11).
CONDILIAC: see BONNOT, É.
COOMBS, C. H., and N. W. SHOCK: Galvanometric reactions of adolescents to olfactory stimuli. Psychol. Bull. **32**, 531 (1935).
COPELMAN, L. S.: Étude de psychologie normale et pathologique du chien. Rev. Pathol gén. physiol. Clin. **57**, 947—954 (1957).
CORNU, F.: Vorläufige Mitteilungen über Arbeiten auf dem Grenzgebiete zwischen Kolloidchemie, Mineralogie und Geologie. V. Der „Tongeruch" und andere Gerüche der Hydrogele des Mineralreichs. Ztschr. Chem. Indust. Kolloide **4**, 90—91 (1909).
CRAIG, W.: The dog as detective. Sci. Mthly **18**, 38—47 (1924).
CRAIGIE, E. H.: A preliminary experiment upon the relation of the olfactory sense to the migration of the sockeye salmon. *(Oncorhynchus nerka,* Walbaum) Trans. roy. Soc. Canada **20**, Sect. V, 215—224 (1926).
CROCKER, E. C.: The rationale of odor classification. J. Soc. Cosmetic Chemists **2**, 15—20 (1950).
—, and F. N. DILLON: Odor directory. Amer. Perf. Ess. Oil Rev. **53**, 297—301, 396—400 (1949).
—, and L. F. HENDERSON: Analysis and classification of odors. An Effort to develop a workable method. Amer. Perfumer **22**, 325—327, 356 (1927).
—, and L. B. SJÖSTRÖM: Odor detection and thresholds. Chem. Eng. News **27**, 1922—1925 (1949).

DAMBACH, C. A., and D. L. LEEDY: Ohio studies with repellent materials with notes on damage to corn by pheasants and other wildlife. J. Wildlife Management 12, 392—398 (1948).
DE DANNE, R.: Die Vererbung der Nasenleistungen und die sich daraus ergebenden Zuchtziele. Berichte. Dritter kynologischer Weltkongreß, 91—94 (1935).
DARWIN, C.: Inherited instinct. Nature, 7, 281 (1873).
— Journal of Researches into the Natural History and Geology of the Countries Visited during the Voyage of H. M. S. Beagle Round the World. 5th Ed. London, Ward, Lock 1889.
— Perception in the lower animals. Nature 7, 360 (1872/73).
DAVIES, J. T.: The mechanism of olfaction. Sympos. Soc. exp. Biol. No. 16, 170—179 (1926).
— L'odeur et la morphologie des molécules. Indust. Parf. 8, 74—79 (1953).
— Olfactory stimulation. Internat. Perfumer 3, 17—22 (1953).
— Olfactory thresholds: a test of a new theory. Perf. Ess. Oil Rec. 46, 15—18 (1955).
— Le seuil olfactif. Mem. Serv. Chim. État 41, 359—361, discussion, 361—362 (1956/57).
— Molecular shape, size and adsorption in olfaction. Proc. II Internat. Cong. Surface Activity 2, 358—369 (1957).
— The rôle of adsorption and molecular morphology in olfaction: the calculation of olfactory thresholds. Biol. Bull. 117, 222—238 (1959).
— A theory of the quality of odours. J. theoret. Biol. 8, 1—7 (1965).
—, and F. H. TAYLOR: A model system for the olfactory membrane. Nature 174, 693—694 (1954).
DEININGER, N., and R. W. MCKINLEY: The design, construction, and use of an odor test room. Amer. Soc. Test. Materials, Spec. Tech. Pub. No. 164, 23—30, Discussion, 30—32 (1954).
— and F. SULLIVAN: Study and evaluation of the odor properties of surfaces. Ann. N. Y. Acad. Sci. 58, 215—224 (1954).
DELANGE, R.: De l'odeur et de ses relations avec la structure moléculaire. Bull. Soc. chim. France [iv] 31, 589—630 (1922).
— Relations entre l'odeur et la constitution chimique des composés organiques. Parfumerie 1, 124—131 (1943).
DERVICHIAN, D.: Du rôle de l'adsorption dans olfaction. Indust. Parf. 4, 137—141; 211—215; 252—256 (1949).
Deutscher Schäferhund-Verband: Beiträge zur Frage der Verwendung von Hunden im Kriminaldienst. Vier Protokolle über Prüfungen der Spürfähigkeit nebst einer Einleitung „Vom Diensthund" von KONRAD MOST.
DEVAUX, H.: Sur l'adsorption des odeurs. Photographies et cinématographie du degagement de parfums et vapeurs diverses. J. Physique Radium 8 D, No. 406 [vii], 85 S—87 S (1937).
DEVAUX, P.: La captation moléculaire des parfums. Indust. Parf. 10, 405, 407—408 (1955).
DILLON, E.: A neglected sense. Nineteenth Century 35, 574—587 (1894).
DIMMICK, F. L.: A note on Henning's smell series. Amer. J. Psychol. 33, 423—425 (1922).
— The investigation of the olfactory qualities. Psychol. Rev. 34, 321—335 (1927).
DÖVING, K. B.: Studies on the responses of bulbar neurons of frog to different odour stimuli. Rev. Laryng. (Bordeaux) 86, 845—854 (1965).
— An electrophysiological study of odour similarities of homologous substances. J. Physiol. 186, 97—109 (1966).

DOLL, W., u. K. BOURNOT: Über den Geruch optischer Antipoden. Pharmazie **4**, 224—227 (1949).
DOROUGH, G. L., H. B. GLASS, T. L. GRESHAM, G. B. MALONE, and E. E. REID: Structure-property relationships in some isomeric octanols. J. Amer. chem. Soc. **63**, 3100—3110 (1941).
DRAVNIEKS, A.: Possible mechanism of olfaction. Nature **194**, 245—247 (1962).
— Contact-potential olfactometry by displacement of adsorbed oriented layers. Electrochem. Soc. May 10, 17 pp. (1962).
— Physicochemical basis of olfaction. Ann. N. Y. Acad. Sci. **116**, 429—439 (1964).
— Odours as signatures. New Scientist **31**, 622—624 (1966).
— Current status of odor theories. Adv. Chem. Ser. **56**, 29—52 (1966).
DUBUISSON, M.: L'olfaction et les ions gazeux. C. R. Ass. franç. Avance. Sci. **43**, 574—577 (1915).
DUMAS, G.: L'odeur de sainteté. Rev. de Paris, 14me Année, 531—552 (1907).
DUMÉRIL, A. A.: Des odeurs, de leur nature et de leur action physiologique. C. R. Acad. Sci. (Paris) **16**, 261—266 (1943).
— Sur les organes des sens et en particulier sur ceux de l'odorat, du goût et de l'ouie dans les poissons. C. R. Acad. Sci. (Paris) **46**, 867—879 (1958).
— Considerations sommaires sur les odeurs. Monit. Sci. **5**, 183—187 (1863). (1935/36).
DUMÉRIL, A. M. C.: Mémoire sur l'odorat des poissons. Mag. Encyclop. **71**, 99—113 (1807).
DUNCAN, R. B., and M. BRIGGS: Treatment of uncomplicated anosmia by vitamin A. Arch. Otolaryngol. **75**, 116—124 (1962).
DURANTON, B.: Radiochimie, radiobiologie et olfaction. Indust. Parf. **2**, 196—198 (1947).
DURRANS, T. H.: The relationship between odour and chemical constitution. Perf. Ess. Oil Rec. **10**, 104—134, 136 (1919).
— Relation of odour to chemical constitution. Perf. Ess. Oil Rec. **10**, 266 (1919).
— The "residual affinity" odour theory. Perf. Ess. Oil Rec. **11**, 391—393 (1920).
— "Odour and constitution". Chemistry and Industry, 41 (1953).
DUSSIK, K. T., u. O. KAUDERS: Untersuchungen zu einer neuen Methodik der Geruchs- und Geschmacksprüfung. Wien. klin. Wschr. **46**, 745—747 (1933).
DYSON, G. M.: Some aspects of the vibration theory of odour. Perf. Ess. Oil Rec. **19**, 456—459 (1928).
— Further aspects of the vibration theory of odour. Perf. Ess. Oil Rec. **20**, 200—201 (1929).
— Raman effect and the concept of odour. Perf. Ess. Oil Rec. **28**, 13—19 (1937).
— The scientific basis of odour. Chemistry and Industry **57**, 647—650, Discussion, 650—651 (1938).
— Odour. Manual of Organic Chemistry **1**, 765—770 (1950). London: Longmans, Green 1950.
— Odour and chemical constitution. Nature **173**, 831 (1954).
EAYRS, J. T., and D. G. MOULTON: Studies in olfactory acuity. I. Measurement of olfactory thresholds in the rat. II. (MOULTON and EAYRS) Relative detectability of *n*-aliphatic alcohols by the rat. Quart. J. exp. Psychol. **12**, 90—98; 99—109 (1960).
EBERS, J. J. H.: Hyperästhesie des Geruchssinnes als forensische Frage. Ein Diebes-Riecher. Vjschr. gerichtl. öffentl. Med. **16**, 278—296 (1859).
EHRENSVÄRD, G.: Über die Primärvorgänge bei Chemozeptoren-Beeinflussung. Acta physiol. scand. **3**, Suppl. IX, 151 pp. (1942).

EISLER, H.: How prothetic is the continuum of smell? Scand. J. Psychol. **4**, 29—32 (1963).

ELSBERG, C. A.: The sense of smell. Introduction. Bull. Neurol. Inst. N. Y. **4**, 1—4 (1935).

— Cerebral activity including conscious sensation as a physicochemical process with evidence to indicate that many vital processes can be expressed in simple mathematical relationships. Bull. Neurol. Inst. N. Y. **7**, 212—244 (1938, 1939).

— and I. LEVY: The sense of smell. I A new and simple method of quantitative olfactometry. Bull. Neurol. Inst. N. Y. **4**, 5—19 (1935/36).

— and H. SPOTNITZ: Are vision and the olfactory sense governed by the same laws? A comparison of quantitive functional tests of vision and of the sense of smell and its significance. Bull. Neurol. Inst. N. Y. **6**, 421—429 (1937).

— E. D. BREWER, and I. LEVY: The sense of smell. III. The relation between the olfactory coefficients and boiling points of odorous substances. Bull. Neurol. Inst. N. Y. **4**, 26—30 (1935/36).

— — — The sense of smell. V. The relative importance of volume and pressure of the impulse for the sensation of smell and the nature of the olfactory process. Bull. Neurol. Inst. N. Y. **4**, 264—269 (1935/36).

— — — The sense of smell. VII. The odorous substances to be used for tests of the olfactory sense. Bull. Neurol. Inst. N. Y. **4**, 286—293 (1935/36).

— I. LEVY, and E. D. BREWER: The sense of smell. II. A new principle for the classification of odors based upon their olfactory coefficients. Bull. Neurol. Inst. N. Y. **4**, 20—25 (1935/36).

ENGEN, T., L. P. LIPSITT, and H. KAYE: Olfactory responses and adaptation in the human neonate. J. comp. Physiol. Psychol. **56**, 73—77 (1963).

— — Decrement and recovery of responses to olfactory stimuli in the human neonate. J. comp. Physiol. Psychol. **59**, 312—316 (1965).

—, and D. H. MCBURNEY: Magnitude and category scales of the pleasantness of odors. J. exp. Psychol. **68**, 435—440 (1964).

—, and C. PFAFFMANN: Absolute judgements of odor intensity. J. exp. Psychol. **58**, 23—26 (1959).

— — Absolute judgments of odor quality. J. exp. Psychol. **59**, 214—219 (1960).

ENGSTRÖM, H., and G. BLOOM: The structure of the olfactory region in man. Acta Oto-Laryng. **43**, 11—21 (1953).

ERB, R. C.: Electrons as a cause of odor. Amer. Perfumer **50**, 561 (1947).

ERDMANN, H.: Über das Verhalten der Geruchsstoffe gegen flüssige Luft. J. prakt. Chem. **169**, 225—228 (1900).

ESTABLE, C.: Systèmes osmatiques et cause histologique possible de la pluralité d'énergies olfactives spécifiques. Trav. Lab. Recherches Biol. Univ. Madrid **22**, 329—358 (1924).

ESTÈVE, L.: Parfums et Belles-Lettres. Paris: Vigot Frères 1939.

EYFERTH, K.: Über die Abhängigkeit der Urteile über olfaktorische Qualitäten von der Empfindung der Reizintensität. Z. exp. angew. Psychol. **12**, 209—222 (1965).

EYKMAN, H. J.: Klinische Olfactometrie. Utrecht: Dissertation 1927.

— Clinische olfactometrie. Ned. Tijdschr. Geneesk. **71**, II, 1678—1679 (1927).

EYSENCK, M. D.: An experimental and statistical study of olfactory preferences. J. exp. Psychol. **34**, 246—252 (1944).

FABANI, C.: L'odorato negli uccelli. Riv. ital. Sci. Nat. **16**, Suppl. Mens. 145—146 (1896).

FABER, Horst Maria: Zärtliches Brevier der Düfte. Wiesbaden: Verlag der Greif. n. d.

FAIR, G. M.: On the determination of odors and tastes in water. J. New Engl. Water Wks. Ass. **47**, 248—269 (1933).

FERRARI, A.: Esquema radio-electromagnético dos 4 sentidos especificos: vista, ouvido, olfato e paladar. Impr. med., Rio de J. **27**, No. 449, 75 (1951).

FERRARI, C., e V. GIULIANI: L'apparato olfattivo e l'organo di Jacobson nella scala zoologica. Ateneo Parmense **28**, 211—242 (1957).

FINDLEY, A. E.: Further studies of Henning's system of olfactory qualities. Amer. J. Psychol. **35**, 436—445 (1924).

FINK, E.: Geruchsorgan und Riechvermögen bei Vögeln. Zool. Jb., physiol. Abt. **71**, 429—450 (1965).

FISCHER, E., and F. PENZOLDT: Über die Empfindlichkeit des Geruchssinnes. S.-B. phys.-med. Soc. Erlangen, Heft 18, 7—10 (1885/86); Ann. Chem. **239**, 131—136 1887).

FLEISSNER, H.: Geruch beim Schlag. Z. angew. Chem. **21**, 1845 (1908).

FLESS, D. A.: Opyty po usileniju obonjanija u sobak. Priroda No. 3, 93—95 (1958).

FORRESTER, A. T., and W. E. PARKINS: A test of the infrared absorption theory of olfaction. Science **114**, 5—6 (1951).

FORTUNATO, V., e P. NICCOLINI: L'Olfatto. Cognizioni Antiche e Recenti. Roma: Istituto Farmacoterapico Italiano 1958.

FOSTER, D.: A quantitative and qualitative analysis of odors. Ithaca, N. Y.: Cornell Univ., Dissertation 1949.

— E. H. SCOFIELD, and K. M. DALLENBACH: An olfactorium. Amer. J. Psychol. **63**, 431—440 (1950).

DE FOURCROY, A. F.: Sur l'esprit recteur de Boerrhaave, l'arome des chimistes français, ou le principe de l'odeur des végétaux. Ann. Chim. **26**, 232—250 (1798).

FREUD, J., and I. E. UYLDERT: Micturation and copulation behaviour patterns in dogs. Acta Brevia Neerland. Physiol. Pharmacol. Microbiol. E. A. **16**, 49—53 (1948).

FREY, S.: Der Polizeihund, seine Geschichte, Zucht und Abrichtung. Berlin: Neufeld und Henius 1911.

FRINGS, H., and W. A. BOYD: Evidence for olfactory discrimination by the bob-white quail. Amer. Midland Naturalist, **48**, 181—184 (1952).

FRISCH, K. v.: Die Bedeutung des Geruchsinnes im Leben der Fische. Naturwiss. **29**, 321—333 (1941).

FRISIUS: Gezicht, gehoor en reuk bij eenden. Nederl. Jager **39**, 40—41 (1933).

FROLOV, IU.: Zaladka obonjanija. Tekhn. Molod. **27**, No. 12, 27—28 and back cover (1959).

FULD, L. F.: The use of police dogs: a summary. J. Amer. Inst. Crim. Law Criminol. **3**, 123—127 [Includes an English translation of the summary of the book by Friedo Schmidt: *Verbrecherspur und Polizeihund]* (1912/13).

FULLER, J. L., S. B. SCHNITZER, and M. E. SCHNITZER: Studies on the tracking behaviour of dogs. I. Breed differences among dogs in locating hidden persons. III. Individual differences in a group of beagles. Armour Research Inst. Chicago and R. B. Jackson Mem. Lab., Bar Harbour, Maine, U.S.A.

FULLMAN, B.: Ability to smell hydrogen canide. Chemistry and Industry 948 (1953).

— Stereochemical theory of olfaction. Nature **199**, 912 (1963).

GALLI, V., and V. INCUTTI: Aspetti emotivi della percezione olfattiva. Arch. ital. Laringol. **72**, 485—492 (1964).

— The history of twins, a criterion of the relative powers of nature and nurture.

GALTON, F.: Arithmetic by smell. Psychol. Rev. **1**, 61—62 (1894).

J. Anthropol. Inst. Gt. Brit. Ireland **5**, 391—406 (1876).

GAMBLE, E. A. McC.: The applicability of Weber's law to smell. Amer. J. Psychol. **10**, 82—142 (1898/99).
— "Der Geruch" by Hans Henning. Amer. J. Psychol. **32**, 290—295 (1921).
GARBINI, A.: Intorno al minimum percettibile di odore. Mem. Accad. Agric. Arte Commercio Verona, [iii] **68**, 87—90 (1893).
— Evoluzione del senso olfattivo nella infanzia. Arch. Antropol. Etnol. **26**, 239—286 (1896).
GATTEFOSSÉ, R. M.: Einstein et les parfums. Parf. Moderne **15**, 253—254 (1922).
GAVAUDAN, P., H. POUSSEL, G. BREBION, et M.-P. SCHÜTZENBERGER: L'étude des conditions thermodynamiques de l'excitation olfactive et les théories de l'olfaction. C. R. Acad. Sci. (Paris) **226**, 1395—1396 (1948).
— — et M.-P. SCHÜTZENBERGER: Étude pharmacodynamique des chimiorécepteurs de l'olfaction. II. L'étude thermodynamique des séries homologues et les théories de l'olfaction. Gallica Biol. Acta **1**, 167—187 (1948).
— G. BREBION, H. POUSSEL, et M.-P. SCHÜTZENBERGER: Étude pharmacodynamique de chimiorécepteurs de l'olfaction. I. Principes et méthodes d'une étude thermodynamique de l'olfaction. Gallica Biol. Acta **1**, 147—166 (1948).
GENOT, R.: "Non mais Docteur, quelle odeur!" Scalpel **114**, 64—68 (1961).
GENZMER, A.: Untersuchungen über die Sinneswahrnehmungen des neugeborenen Menschen. Halle: Dissertation 1873.
GERARDIN: Dosage des odeurs. Rev. Hyg. **17**, 597—600 (1895).
GERARDIN, A., et M. NICLOUX: Mesure des odeurs de l'air. C. R. Acad. Sci (Paris) **122**, 954—955 (1896).
GEREBTZOFF, M. A.: L'olfaction: structure de l'organe olfactif et mécanisme de l'olfaction. J. Physiol. **45**, 247—283 (1953).
— Les théories modernes de l'olfaction. Françe Parf. **6**, 316—320 (1963).
— et E. PHILIPPOT: Lipides et pigment olfactifs. Acta oto-rhino-laryng. Belg. **11**, 297—300 (1957).
Gesellschaft für Hundeforschung: Bericht über die erste Haupttagung der Gesellschaft in Berlin und Umgebung vom 15. bis 18. 9. 1931. Z. Hundeforsch. **2**, 3—11 (1932).
— Bericht über die Umfrage der Gesellschaft vom Jahre 1931. Z. Hundeforsch. **2**, 126—129 (1932).
GESTELAND, R. C.: Initial events of the electro-olfactogram. Ann. N. Y. Acad. Sci. **116**, 440—447 (1964).
— The mechanics of smell. Discovery **27**, No. 2, 29—34 (1966).
GHELARDI PRIMAVORI, G.: Il muco nasale come probabile attualizzatore dell'odore potenziale di alcuni composti. Boll. Soc. Ital. Biol. sper. **30**, 266—267 (1954).
GIANNUZZI, J.: Recherches physiologiques sur les nerfs de l'olfaction. C. R. Soc. Biol. **5**, 97—100 (1863).
GIESSLER, C. M.: Wegweiser zu einer Psychologie des Geruches. Hamburg und Leipzig: Voss 1894.
— Die Gerüche vom psychogenetischen Standpunkte aus. Vjschr. wiss. Philos. **26**, 49—76 (1902).
GIETZ, F.: Wie kommt das Geruchserlebnis zustande? Pharmazie, **7**, 489—495 (1952).
— Ist das Duften der Riechstoffe durch infrarote Strahlen bedingt? Ber. Schimmel & Co., Leipzig-Miltitz 135 (1952/53).
GILL, E. L.: The condor's sense of smell. Trans. Nat. Hist. Soc. Northumberland, Durham, Newcastle-upon-Tyne **1**, 40 (1904/07).
GILLESPIE, T. H.: Taste or scent in the ostrich. Scot. Naturalist Nos. 129—130; 168 (1922).
— Do birds posses a sense of smell? Scotsman, Jan. 14 (1956).

GLOAG, J.: The New Pleasure. London: Allen & Unwin 1933.
GODINEAU, H.-F.: Considérations sur l'Odorat. Paris: Didot Jeune 1809.
GÖZ, H.: Über den Art- und Individualgeruch bei Fischen. Z. vgl. Physiol. **29**, 1—45 (1941/42).
GONNET, P. J.: Le Miel. Parfum-Saveur Arôme. Montfavet, Vaucluse: Librairie de Vulgarisation Apicole 1951.
GORDON, K.: The recollection of pleasant and unpleasant odors. J. exp. Psychol. **8**, 225—239 (1925).
GREELEY, W. R.: A quantitative study on the perceptive function of the olfactory organ. New Engl. J. Med. **264**, 1302 (1961).
GRIESBACH, H.: Vergleichende Untersuchungen über die Sinnesschärfe Blinder und Sehender. Arch. ges. Physiol. **74**, 577—638; **75**, 365—429; 523—573 (1899).
GRIJNS, G.: Messungen der Riechschärfe bei Europäern und Javanen. Arch. Physiol. 509—517 (1906).
— Is er verband tusschen het opslorpend vermogen voor stralende warmte en de riekkracht van stoff? Vers. Gewone Vergad. Wis. Natuurk. Afd. K. Akad. Wetensch. Amsterdam **27**, 280—282 (1918).
— Y-a-t-il une relation entre le pouvoir absorbant à l'égard de la chaleur rayonnante et le pouvoir odorant des substances. Arch. néerland. Physiol. **3**, 377—390 (1919).
GROEBBELS, F.: Der Vogel. I. Atmungswelt und Nahrungswelt. Berlin: Borntraeger 1932.
GRUCH, W.: Über die Riechfähigkeit bei Wanderratten. Zool. Jahrb., Abt. allg. Zool. Physiol. Tiere **67**, 65—80 (1957/58).
GRUHN, C.: Der Geruch der Metalle. Dtsche Uhrmacher-Ztg. **31**, 210 (1907).
GÜTTICH, H.: Über die klinische Bedeutung der Riechprüfung. Mschr. Ohrenheilk. Laryngo-Rhinol. **98**, 49—53 (1964).
— Intravenöse verabreichte Riechstoffe: Gustatorisches Riechen. HNO **13**, 42—45 (1965).
GUGLIELMINOTTI, L.: L'influenza del trascendentalismo relativista sulle teorie chimico-fisiche dell'olfazione. Profumi Italici **3**, 237—242 (1925).
GUILLOT, M.: Anosmies partielles et odeurs fondamentales. C. R. Acad. Sci. (Paris) **226**, 1307—1309 (1948).
— Sur la relation entre l'odeur et la structure moléculaire. C. R. Acad. Sci. (Paris) **226**, 1472—1474 (1948).
— Hypothèse de l'nfluence d'impuretés isostériques sur l'odeur des composés organiques. C. R. Acad. Sci. (Paris) **227**, 468—470 (1948).
— Structure moléculaire et odeur. Indust. Parf. **3**, 380—384 (1948).
— Rotatory power and olfaction. Recherches [English Ed.], No. 5, 24—31 (1955).
— Sur les mécanismes psycho-physiologiques de l'olfaction. J. Psychol. norm. Pathol. **55**, 1—20 (1958).
— Sur quelques aspects physiques et psychiques des parfums. France Parf. **6**, 123—130 (1963).
— Sur la discrimination des substances odorantes. Recherches, No. 14, 4—11 (1964).
— et R. BABIN: Pouvoir rotatoire et odeur. C. R. Acad. Sci. (Paris) **229**, 1363—1365 (1949).
— et A. BERTON: Sensibilité comparée des détecteurs physique et physiologique des vapeurs odorantes. France Parf. **6**, 403—407 (1963).
— et A. FIEHRER: Folliculine et olfaction. C. R. Soc. Biol. **143**, 922—923 (1949).
— et Madame M. GUILLOT: Sur le comportement électrique des vapeurs organiques odorantes dans l'air. C. R. Acad. Sci. (Paris) **224**, 39—41 (1947).
— et S. GUILLOT-ALLEGRE: Sur le comportement des vapeurs odorantes dans l'air. Recherches, No. 1, 4—18 (1952).

— et F. Picon: Sur l'odeur des produits chimiques définis purs. Recherches, No. 13, 104—110 (1963).
Guillot, M., et P. Thibaut: Influence du pouvoir rotatoire sur la qualité de l'odeur. C. R. Acad. Sci. (Paris) **232,** 1138—1140 (1951).
— Sur la psychophysiologie des odeurs. Rev. Laryng. (Bordeaux) **86,** 959—971 (1965).
— Examen des différents facteurs qui conditionnent l'appréciation d'un arôme par un sujet humain normal. France Parf. **9,** 139—146 (1966).
Gurney, J. H.: On the sense of smell possessed by birds. Ibis [xi] **4,** 225—253 (1922).
Gurney, P. L. C.: Police dog training: German practice. Police J. **3,** 526—534 (1930).
Gwilt, J. R.: Odour of potassium cyanide. Lancet **II,** 1197 (1960).
— The odour of (potassium) cyanide. Medico-Legal J. **29,** 98—99 (1961).
Hainer, R. M.: Some suggested critical experiments in olfactory theory. Ann. N. Y. Acad. Sci. **116,** 477—481 (1964).
— A. G. Emslie, and A. Jacobson: An information theory of olfaction. Ann. N. Y. Acad. Sci. **58,** 158—174 (1954).
Haldane, J. B. S.: Physiological properties of some common gases at high pressures. Chem. Products **4,** 83—84 (1941).
Haler, D.: Does iron smell? Lancet **267,** 143 (1954).
Halla, F., u. A. Schmölzer: Geruch und Gesteinsfunken. Naturwiss. **32,** 217 (1944).
—, u. R. van Tassel: Geruch von Gesteinsfunken. Naturwiss. **43,** 444 (1956).
Haller, A. v.: Olfactus. Elementa Physiologiae Corporis Humani. Liber XIV. Tomus Quintus. Pp. 125—185. Lausannae: Francisci Grasset 1763.
Halliday, J. L.: A reconsideration of the sense of smell. Glasgow med. J. **105,** 182—199 (1926).
Hamilton, G. van T.: An experimental study of an unusual type of reaction in a dog. J. comp. Neurol. **17,** 329—341 (1907).
Hamrum, C. L.: Experiments on the senses of taste and smell in the bob-white quail *(Colinus virginianus).* Amer. Midland Naturalist **49,** 872—877 (1953).
Hansen, D.: Eine einfache Methode zur quantitativen Geruchsprüfung in Klinik und Praxis. Arch. Ohren-, Nasen-, Kehlkopfheilk. **185,** 760—763 (1965).
Hansmann, J.: Die Physiologie des Riechens in ihrer Bedeutung für die Nasenarbeit des Hundes. Ztg. Ver. dtsch. Schäferhunde **28,** 736—741 (1929).
— Unter welchen Gesichtspunkten erfolgt die praktische Verwendung des Polizeifährtenhundes? Z. Hundeforsch. **1,** 14—30 (1931).
— Der Gebrauchshund auf der Menschenfährte. Unsere Hunde **3,** 275—278, 280—281 (1949); Unser Rassehund **2,** 334—336, 385—386, 430—431 (1950); **3,** 84, 114, 129—130 (1951).
— Hunde- und andere Nasen, Geruch und „Roter Faden". Terrier **48,** 206—207, 210—212 (1955).
Hara, T. J., K. Ueda, and A. Gorbman: Electroencephalographic studies of homing salmon. Science **149,** 884—885 (1965).
Harper, R., E. C. Bate-Smith, and D. G. Land: Odour description and odour classification. Norwich: Food Research Inst. 1966.
— On odour classification. J. Food Technol. **1,** 167—176 (1966).
—, E. C. Bate-Smith, and D. G. Land: Odour description and odour classification. A multidisciplinary examination. London: Churchill 1968.
Harvey, H. G.: Survey of odour in packaging of foods. London: Inst. of Packaging 1964.
Hasama, B.: Über die elektrischen Begleiterscheinungen an der Riechsphäre bei der Geruchsempfindung. Arch. ges. Physiol. **234,** 748—755 (1934).

Hasler, A. D.: Odour perception and orientation in fishes. J. Fish. Res. Board Canada **11**, 107—129 (1954).
— Perception of pathways by fishes in migration. Quart. Rev. Biol. **31**, 200 (1956).
—, and W. J. Wisby: Use of fish for the olfactory assay of pollutants (phenols) in water. Trans. Amer. Fish. Soc. **79**, 64—70 (1950).
— — Discrimination of stream odors by fishes and its relation to parent stream behavior. Amer. Naturalist **85**, 223—238 (1951).
Hattinger, A.: Skripten für die Ausbildung von Gendarmerie-Lawinensuchhunden. Heft 3. Wien: Bundesministerium für Inneres 1958.
Hawkes, L.: Fragments of jet coal in the chalk of Kent. Proc. Geologists' Assoc. **56**, 1—7 (1945). Discussion, 7—9.
Hayashi, T. (Editor): Olfaction and taste. II. Oxford, London, Edinburgh, New York, Toronto, Sydney, Paris. Braunschweig: Pergamon Press 1967.
Hazzard, F. W.: A descriptive account of odors. J. exp. Physiol. **13**, 297 (1930).
Hearn, L.: In Ghostly Japan. London: Kegan, Paul 1905.
Heinroth, O., u. M. Heinroth: Die Vögel Mitteleuropas. Berlin-Lichterfelde: Hugo Bermühler Verlag n. d.
Heitzenroeder, C.: Über das Verhalten des Hundes gegen einige Riechstoffe. Z. Biol. **62**, 491—507 (1913).
Heller, H.: Duft und Geruch. Prometheus **29**, 418—419, 426—428 (1917/18).
— Zur Geruchstheorie von Teudt. I. II. Prometheus **30**, 396—398 (1918/19); **31**, 355—356 (1919/20).
— Über die Geruchstheorie von Teudt. Biol. Zbl. **39**, 364—370 (1919); **41**, 138—142 (1921).
— Duft und Geruch. Dtsch. Parf.-Ztg. **5**, 2—3, 12—13 (1919).
— Der Duft als analytisches Kennzeichen. Dtsch. Parf.-Ztg. **5**, 91—92 (1919).
— Über eine neue Theorie von Duft und Geruch. Dtsch. Parf.-Ztg. **5**, 239—241 (1919).
— Düfte im Mineralreich. Dtsch. Parf.-Ztg. **6**, 60—62, 245—247 (1920).
Hennebert, P. E.: L'olfaction. Acta oto-rhino-laryng. Belg. **7**, 103—225 (1953).
— Geruchsspiele in Japan. Ein Beitrag zur Psychologie des Japaners, des Spiels und der Zeremonie. Z. angew. Psychol. **14**, 322—332 (1919).
— Geruchsversuche am Hund. Z. Biol. **70**, 1—8 (1919/20).
Henning, H.: Der Geruch. Leipzig: Barth. 1. Aufl. 1916; 2. Aufl. 1924.
Henton, W. W., J. C. Smith, and D. Tucker: Odor discrimination in pigeons. Science **153**, 1138—1139 (1966).
Hermanides, J.: Over de Constanten der in de Olfactometrie Gebruiklijke negen Standaardgeuren. Utrecht: Oosthoek 1909.
Herrmann, J.: Gesamterlebnisse bei Gerüchen. Neue psychol. Studien 1, 477—506. Komplexqualitäten, Gestalten und Gefühle. München C. H. Beck'sche Verlagsbuchhandlung (1926).
Hesse, W.: Bestimmung von Geruchsschwellen in absoluten Werten. Z. Hals-, Nasen-, Ohrenheilk. **16**, 359—373 (1926).
— Ein einfacher Apparat zur Bestimmung von Geruchsschwellen in absoluten Werten. Z. Hals-, Nasen-, Ohrenheilk. **19**, 348—352 (1927).
Heusghem, C., et M. A. Gerebtzoff: Résultats concordants des analyses biochimiques et histochimiques des lipides de la muqueuse olfactive. C. R. Soc. Biol. **147**, 540—541 (1953).
Heyer, F.: Über Lawinensuchhunde — Märchen und Wirklichkeit. Terrier **48**, 54—56 (1955).
Heyl, G. E.: Process for treating seeds, bulbs, tubers, and roots. U. S. Patent Aug. 8, No. 2, 168, 523 (1939).

HEYNINX, A.: Essai d'olfactique physiologique. Bruxelles: Larcier 1919.
— La physiologie de l'olfaction. Rev. Oto-Neuro-Ophthalmol. **11**, 10—19 (1933).
HILL, A.: Can birds smell? Nature **71**, 318—319 (1904/05).
— Vapour density and smell. Nature **80**, 427—428 (1909).
HIRAIZUMI, T.: The ancient art of incense in Japan. Soap, Perfumery, Cosmetics **38**, 976—981 (1965).
HOEFT, A.: Untersuchungen über ein Verfahren zum Messen der Riechschärfe von Arbeitshunden. Berlin: Dissertation 1924.
HOFMANN, F. B.: Zur Theorie des Geruchsinnes. I. Parosmie-Studien. II. Über sogenannte Geruchsermüdung. Z. Biol. **73**, 29—66 (1921); **78**, 63—90 (1923).
HOLMAN, A.: My dog rex. The story of police dog rex III. London: Harrap 1957.
HOLWAY, A. H., R. M. STUART, C. A. WINCHELL, and M. J. ZIGLER: On the psychophysics and neurophysiology of olfaction. J. gen. Psychol. **23**, 65—73 (1940).
HONGÔ, N.: Die experimentellen Untersuchungen über den Geruchssinn. I. Über die Veränderungen der Atembewegung bei der Olfaktoriusreizung und über eine neue olfaktometrische Methode beim Kaninchen. II. Über Veränderungen des Geruchsvermögens des Kaninchens bei den akuten und chronischen Vergiftungen von verschiedenen Arzneimitteln. III. Über die Veränderungen des Geruchsvermögens des Kaninchens bei verschiedenen Ernährungsstörungen. IV. Über die Veränderungen des Geruchsvermögens des Kaninchens bei Entzündungsprozessen. Tohoku Med. J. **19**, Suppl. 1, 213—256, Suppl. 3, 91—149, 150—191, 192—215 (1936).
HORNBOSTEL, E. M. v.: Über Geruchshelligkeit. Arch. ges. Physiol. **227**, 517—538 (1931).
HOXIE, W.: The sense of smell in the American vultures. Ornithol. Ool. **12**, 61 (1887).
— The sense of smell in the black vulture. Ornithol. Ool. **12**, 132 (1887).
HSÜ, E. H.: A factorial analysis of olfaction. Psychometrica **11**, 31—42 (1946).
HUIJER, C.: De Olfactologie van aniline en homologen. Utrecht: Dissertation 1917.
HUKUI, H.: Die Bedeutung der physikalisch-chemischen Eigenschaften in der Geruchsphysiologie. Jap. J. Med. Sci., III. Biophysics **2**, 87—88 (1931/33).
— Eine neue odoroskopische Methode zur Prüfung riechender Stoffe in der Luft. Jap. J. Med. Sci., III. Biophysics. **2**, 157 (1931/33).
HUNTER, P.: Powers of smell ascribed to the vulture. Mag. Nat. Hist. **3**, 449 (1830).
— The means by which the vulture *(Vultur Aura* L.) traces its food. Mag. Nat. Hist. **6**, 83—88 (1833).
HUTTON, J.: Nest of the bengal vulture *(Vultur Bengalensis)* with observations on the power of scent ascribed to the vulture tribe. J. Asiatic Soc. Bengal **6**, 112—118 (1837).
HUXLEY, T. H.: On sensation and the unity of structure of sensiferous organs. Nineteenth Century **5**, 597—611 (1879).
HUYGHEBAERT, L.: Le chien. Psychologie, olfaction, mécanisme de l'odorat. Bruxelles: Lebègue n. d.
— Die Vererbung der Nasenleistungen und die sich daraus ergebenden Zuchtziele. Berichte. Dritter kynologischer Weltkongreß 86—87 (1935).
ICHIHARA, M.: Kyukaku no rinsho. Tokyo: Kanehara 1963.
IDE, M.: L'odorat du chien. Rev. Méd. Louvain 271—272 (1926).
IDLER, D. R., U. H. M. FAGERLUND, and H. MAYOH: Olfactory perception in migrating salmon. I. *L*-Serine, a salmon repellent in mammalian skin. J. Gen. Physiol. **39**, 889—892 (1956).

IDLER, D. R., J. R. McBRIDE, R. E. E. JONAS, and N. TOMLINSON: Olfactory perception in migrating salmon. II. Studies on a laboratory bio-assay for homestream water and mammalian repellent. Canad. J. Biochem. Physiol. **39**, 1575—1584 (1961).

ISHIHARA, K.: Zur Kenntnis des Nasenhöhlenorgans der Vögel. Ztschr. Anat. Entwickl. Ges. **98**, 548—577 (1932).

JACKSON, R. T.: Chemical properties of the olfactory tissue. Diss. Absts. **20**, 4425—4426 (1960).

— The olfactory pigment. J. cell. comp. Physiol. **55**, 143—147 (1960).

JAEGER, G.: Über die Bedeutung des Geschmacks- und Geruchsstoffes. Z. wiss. Zool. **27**, 319—331 (1876).

— Entdeckung der Seele. Kosmos, Z. einheitl. Weltanschau. Grund Entwicklungslehre **4**, 171—191 (1878/79).

— Improvements in the preparation of food and other articles for medicinal purposes. Brit. Patent No. 6001, 1884.

— Lehrbuch der allgemeinen Zoologie. III. Abt., Entdeckung der Seele. 3. Aufl. Leipzig: Ernst Günthers Verlag 1884.

JELLINEK, J. S.: The physico-chemical behavior of perfume materials in various carriers. Amer. Perf. Aromat. **73**, 27—31 (1959).

JELLINEK, P.: Die psychologischen Grundlagen der Parfümerie. Heidelberg: Hüthig 1951.

JEROME, E. A.: Olfactory thresholds measured in terms of stimulus pressure and volume. Arch. Psychol. No. 274, 44 pp. (1942).

JOHNSEN, A.: Über die Funken und den Geruch beim Aneinanderschlagen von Mineralien. Naturwiss. **7**, 459—461 (1919).

JOHNSON, H. M.: A note on the supposed olfactory hunting-responses of the dog. J. Animal Behav. **4**, 76—78 (1914).

JOHNSTON, J. W.: Infrared loss theory of olfaction untenable. Physiol. Zool. **36**, 266—273 (1953).

JOHNSTON, J. W., jr.: An application of the steric odor theory. Georgetown med. Proc. sci. Sect. Toilet Goods Ass. No. 34, 4—7 (1960).

— The specificities of human olfaction. Bregman and Dravnieks: Surface effects in detection. Washington: Spartan Books Inc. 1965.

— Judgements of odorant similarity and dissimilarity. Proc. Sci. Sect. Toilet Goods Ass. No. 44, 32—41 (1965).

— An application of the steric odor theory. Georgetown Med. Bull. **17**, 40—42 (1963).

—, and A. B. PARKS: Odor-intensity and the stereochemical theory of olfaction.

—, and A. SANDOVAL: Organoleptic quality and the stereochemical theory of olfaction. Proc. sci. Sect. Toilet Goods Ass. No. 33, 3—9 (1960).

— — The stereochemical theory of olfaction. IV. The validity of muskiness as a primary odor. Proc. sci. Sect. Toilet Goods Ass. No. 37, Suppl. 34—35 (1962).

JONES, F. N.: A test of the validity of the Elsberg method of olfactometry. Amer. J. Psychol. **66**, 81—85 (1953).

— Olfactory thresholds in the international critical tables. Science **118**, 333 (1953).

— An olfactometer permitting stimulus specification in molar terms. Amer. J. Psychol. **67**, 147—151 (1954).

— A comparison of the methods of olfactory stimulation: blasting *vs* sniffing. Amer. J. Psychol. **68**, 486—488 (1955).

—, and M. H. JONES: Modern theories of olfaction: a critical review. J. Psychol. **36**, 207—241 (1953).

JONES, I. H.: The Elsberg precision measurements of the sense of smell. Trans. Pacif. Cst. oto-ophthal. Soc. **25**, 197—216 (1940).
JONGENS, G. W.: Hebben vogels reuk? Levende Natuur **43**, 127—128 (1938/39).
JONGKEES, L. B. W.: Olfactory nystagmus. Acta oto-laryngol. **44**, 556—557; Discussion 557—559 (1954).
JUHÁSZ, A.: A szagerzetek egý uj tulajdonsaga. Athenaeum (Budapest) **12**, 34—39 (1926); English summary 90—91.
— Über eine neue Eigenschaft der Geruchsempfindungen. Ber. IX. Kongress exp. Psychol., München, April 1925, 178—179 (1926).
— Zur Grundlegung der Geruchspsychologie. Proc. Papers 8th Internat. Congress Psychol., Groningen 1926, 305—309 (1927).
JUHÁSZ, K.: Über das Zusammenarbeiten von Geruchs- und Geschmackssinn. Dtsch. Parf.-Ztg. **28**, 217—220 (1942).
KALININA, T. E.: O vlijanii zapachovych razdrazenij na vyssuju nervnuju dejaelnost sobak. Zhurn. Vyss. Nerv. Deyatel **11**, 318—321 (1961); English 330—333.
KALISCHER, O.: Weitere Mitteilung über die Ergebnisse der Dressur als physiologischer Untersuchungsmethode auf den Gebieten des Gehör-, Geruchs- und Farbensinns. Arch. Anat. Physiol., physiol. Abt. 303—322 (1909).
KALMUS, H.: The discrimination by the nose of the dog of individual human odours and in particular of the odours of twins. Brit. J. Animal Behav. **3**, 25—31 (1955).
—, and S. J. HUBBARD: The chemical senses in health and disease. Springfield, Ill.: Thomas 1960.
KATZ, S. H., and V. C. ALLISON: Stenches for detecting leakage of blue water-gas and natural gas. U. S. Bur. Mines, Tech. Papers. No. 267, 22 pp. (1920).
KATSUMATA, S.: Gleams from Japan. Tôkyô: Kenkyûsha 1937.
KELLER, H.: The world I live in. London: Hodder & Stoughton 1908.
KENNEL, P.: Essai de classification des odeurs par la methode de majorités. Arch. Psychol. **15**, 375—378 (1915).
KENNETH, J. H.: Osmics. The science of smell. Edinburgh: Oliver & Boyd. No. 1, 1922; No. 2, 1924.
— Relation of language to physiological stimuli. Nature **116**, 748—749 (1925).
— An experimental study of affects and associations due to certain odours Psychol. Monographs **37**, No. 2, 64 pp. (1927).
— Note on a forgotten sixteenth century disputation on smell. J. Laryngol. Otol. **43**, 103—104 (1928).
— A bibliography of osmics. Perf. Ess. Oil Rec., Year book diary, xiv—xx, xxii, xxiv, xxvi, xxviii (1928).
KETTER, C.: Neue methodisch verbesserte Untersuchungen zur weiteren Klärung eines Wettereinflusses auf die Geruchsempfindung. München: Dissertation 1925.
KIDO, M., and M. YOSHIDA: On the common properties of olfaction. Jap. J. Psychol. **12**, 95—111 (1937).
KING, J. E., R. F. BECKER, and J. E. MARKEE: Studies on olfactory discrimination in dogs. III. Ability to detect human odour trace. J. Animal. Behav. **12**, 311—315 (1964).
KIRWAN, H. A. P. A.: De l'odorat et de l'influence des odeurs sur l'économie animale. Paris: Didot Jeune 1808.
KISTIAKOWSKY, G. B.: On the theory of odors. Science **112**, 154—155 (1950).
KLAGES, W., u. I. KLAGES: Zur Psychologie und Psychopathologie des Geruchsinnes. Arch. Psychiat. Nervenkrankh. **205**, 37—48 (1964).
KLEERKOPER, H., and J. A. MOGENSEN: The chemical composition of scent of fresh water fish with special reference to amines and amino-acids. Z. vgl. Physiol. **42**, 492—500 (1959).

— Role of olfaction in the orientation of *Petromyzon marinus*. I. Response to a single amine in prey's body odor. Physiol. Zool. **36**, 347—360 (1963).
KLOEK, J.: The smell of some steroid sex-hormones and their metabolites. Reflections and experiments concerning the significance of smell for the mutual relation of the sexes. Psychiat. Neurol. Neurochir. **64**, 309—344 (1961).
— De betekenis van de reukzin voor het seksuelle leven; een experimenteel onderzoek. Ned. T. Geneesk. **106**, 2204—2208 (1962).
KLOTZ, L.-P.: L'évolution de l'olfactométrie. Hist. Méd. **6**, No. 2, 35—37; 39—41; 43—45; 47—49; 51—55 (1956).
KLOUWEN, M. H.: Chemical structure and odour. Perf. Ess. Oil Rec., Special No., xxvii—xxxiii (1959).
— Chemische Konstitution und Geruch. II. Homologe Reihen und ihre Gerüche. Parf. Kosmet. **41**, 178—179 (1960).
—, et A. H. RUYS: Constitution chimique et odeur. III. Comparaison entre les propriétés olfactives de quelques benzaldéhydes substitués et celles des nitrobenzènes substitués de la même manière. Parf. Cosmét. Savons **6**, 6—12 (1963).
— Chemische Konstitution und Geruch. III. Vergleich der Geruchseigenschaften einiger Benzaldehyde und Nitrobenzole mit jeweils gleichen Substituenten. Parf. Kosmet. **43**, 289—292 (1962).
— J. G. J. KOK, and RUYS: Chemical constitution and odour. IV. Synthesis and olfactive properties of 11-oxahexadecanolide. J. Soc. Cosmetic Chemists **14**, 19—25 (1963); Discussion 25—28.
KOPPLIN, J. O., J. R. EATON, and J. E. CHRISTIAN: Studies on the adsorption of odorous materials. I. Surface potential changes of solid and liquid adsorbing surfaces. J. Amer. pharm. Ass. **48**, 427—430 (1959).
— II. Surface potential changes due to the adsorption of alcohol vapors. J. Amer. pharm. Ass. **48**, 521—525 (1959).
KOSTER, E. P.: Adaptation, recovery and specificity of olfactory receptors. Rev. Laryng. (Bordeaux) **86**, 880—894 (1965).
KOSTER, S.: Eeen vergelijkend reukonderzoek bij Inlanders van Ned.-Oost Indië en Hollanders. Geneesk. T. Ned. Ind. **61**, 165—189 (1921).
KRACK: Die geruchsphysiologischen und psychologischen Voraussetzungen für die Verwendung des Hundes im Kriminaldienst. Kriminal. Monatshefte **3**, 79—81 (1929).
KRALL, K.: Versuche mit dem Klugen Hans. II. Prüfung des Geruchsinnes. Tierseele **1**, 208—214 (1914).
KRAVKOV, S. V.: The influence of odors upon color vision. Acta ophthalmol. **17**, 426—442 (1939).
KREMER, J. H.: Adsorption de matières odorantes et de narcotiques odorants par les lipoïdes. Arch. néerl. Physiol. **1**, 715—725 (1917).
KRISCH, W.: Eine neue Theorie der Sinnesorgane. Naturwiss. **21**, 876—879 (1933).
KRUMM-HELLER, A.: Vom Weihrauch zur Osmotherapie. Geschichte und Beiträge zu einer Riechstoff-Heilkunde. Berlin-Steglitz: Becker 1934.
— Del incienso a la osmoterapia. Barcelona: Librería Sintes 1940.
KRUSHINSKII, L. V., A. V. CHUVAEV u N. YU. VOLKIND: Novye dannye po izucheniiu chut'ia u sobak. Zool. Zhurn. **25**, 373—382 (1946).
—, and D. A. FLESS: Metodika primineniia broma i kofeina pri podgotovke rezyshknykh sobak. Moskva: Dissertation 1956.
— — Iskusstvennoe usilenie obonjanija u služebnych sobak. Zhurn. Vyss. Nervn. Deiatel. **9**, 284—290 (1959).
—, —, and V. V. RILOV: Cutje sobak i ego rolj v poiske po sledu. Bjull. Moskovsk. Obshest. Ispitt. Prirody: Otdel. Biol. **65**, No. 1, 127—134 (1960).

KUBO, I.: Kyukakuno seiri oyobi shinri. Tokei Shimpo No. 3, 5—12 (1910).
KUDRIAVTZEV, A. A., and N. I. LOZHKIN: K metodike izucheniya obonianija u korov. Fiziol. Zhurn. SSSR **42**, 916—918 (1956).
KUNAYEFF, A.: Izsliedovaniye oboniatelnoi funktsii u sobaki po metodu sochetatelno-dvigatelnikh refleksov. Obozr. Psikhiat. Nevrol. Eksp. Psikhol. **17**, 215—232 (1912).
KUSAKA, K., and T. MORI: A study of tracking and discriminating scents by the police dog. (1962).
KUSSMAUL, A.: Untersuchungen über das Seelenleben des neugeborenen Menschen. Leipzig und Heidelberg: C. F. Winter'sche Verlagsbuchhandlung 1859.
LAAS, D.: Die Physiologie und Psychologie des Geruches. Seifen-Öle-Fette-Wachse **75**, 268—271, 289—291 (1949).
LAFARGUE, P.: The rise and fall of Dr. Vaughan-Trotter. Practitioner **61**, 609—617 (1898).
LAFFORT, P.: Essai de standardisation des seuils olfactifs humains pour 192 corps purs. Arch. sci. Physiol. **17**, 75—105 (1963).
— Mise en évidence de relations linéaires entre l'activité odorante des molécules et certaines de leurs caracteristiques physicochimiques. C. R. Acad. Sci. (Paris) **256**, 5618—5621 (1963).
— Efficacité odorante et activité thermodynamique. Rev. Laryng. (Bordeaux) **86**, 860—879 (1965).
LAIRD, D. A.: How the consumer estimates quality by subconscious sensory impressions with special reference to the role of smell. J. appl. Psychol. **16**, 241—246 (1932).
LANCUM, F. H.: Home life of the badger. Agriculture **59**, 435—440 (1952/53).
LANGLOIS, G.: Influence sur l'odeur de l'existence de cycles dans les molécules organiques. Parfumerie **1**, 239—240 (1943).
LARGUIER DES BANCELS, L.: Le goût et l'odorat. Arch. Psychol. **10**, 1—46 (1910).
LASSWITZ, K.: Bilder aus der Zukunft. Zwei Erzählungen aus dem vierundzwanzigsten und neunundzwanzigsten Jahrhundert. Breslau: Schottlaender 1878.
LAUFFER, P. G.: Odor and olfaction. II. Olfaction's place among biological processes. Drug and Cosmet. Indust. **67**, 326—327, 410—417 (1950).
— Olfaction and cholinesterase. Amer. Perf. Aromat. **74**, No. 2, 36—40, 42, 44 (1959).
LE GALLIENNE, R.: The romance of perfume. New York and Paris: R. Hudnut 1928.
LEICHER, H.: Sättigungsdrücke und Dampfdichte verschiedener Riechstoffe (zum Gebrauch für quantitative Geruchsprüfungen). Z. Laryngol. Rhinol. Otol. Grenzgeb. **18**, 222—226 (1929).
LEIGHTON, A. H.: The turkey vulture's eyes. Auk N. S. **45**, 352—355 (1928).
LEITCH, N.: The science of smells. Middlesex Hosp. J. **34**, 166—183 (1934).
LE MAGNEN, J.: Étude d'une méthode d'analyse qualitative de l'olfaction. Ann. psychol. **43—44**, 249—264 (1942/43).
— Étude des facteurs dynamiques de l'excitation olfactive. Ann. psychol. **45—46**, 77—89 (1944/45).
— Un cas de sensibilité olfactive se présentant comme un caractère sexuel secondaire feminin. C. R. Acad. Sci. (Paris) **226**, 694—695 (1948).
— Analyse d'odeurs complexes et homologues par fatigue. C. R. Acad. Sci. (Paris) **226**, 753—754 (1948).
— Caractérisation physique et physiologique des substances odorantes. Indust. Parf. **3**, 170—174 (1948).
— Variations spécifiques des seuils olfactifs chez l'homme sous actions androgène et oestrogène. C. R. Acad. Sci. (Paris) **228**, 947—948 (1949).

LE MAGNEN, J.: Aperçu sur l'intervention des phénomènes de surface dans l'excitation olfactive. Parf. moderne **41**, No. 14, 30—33 (1949).
— Odeurs et Parfums. Paris: Presses Univ. de France 1949.
— La Variation des sensibilités olfactives en fonction d'états physiologiques généraux et le mécanisme de discrimination des qualités olfactives. Paris: Dissertation 1949.
— Nouvelles données sur le phénomène de l'exaltolide. C. R. Acad. Sci. (Paris) **230**, 1103—1105 (1950).
— Les phénomènes olfacto-sexuels chez l'homme. Arch. sci. Physiol. **6**, 125—160 (1952).
— L'Olfaction: le fonçtionnement olfactif et son intervention dans les régulations psycho-physiologique. J. Physiol. **45**, 285—326 (1953).
— Quelques aspects généraux de la physiologie de l'olfaction. France Parfums **6**, 29—38 (1963).
— L'analyse discriminative de l'appareil olfactif. Rev. Laryngol.-Otol.-Rhinol. **86**, 827—831 (1965).
—, et A. RAPAPORT: Essai de détermination du rôle de la vitamine A dans le mécanisme de l'olfaction chez le rat blanc. C. R. Soc. Biol. **145**, 800—803 (1951).
LETTVIN, J. Y., and R. C. GESTELAND: Speculations on smell. Cold Spr. Harb. Symp. quant. Biol. 30, 217—225 (1965).
LEWIS, J. B.: Sight and scent in the turkey vulture. Auk N. S. **45**, 467—470 (1928).
LIEGEOLS: Mémoire sur les mouvements de certains corps organiques à la surface de l'eau et sur les applications qu'on peut faire à la théorie des odeurs. Arch. Physiol. norm. Pathol. **1**, 35—55, 236—257 (1868).
LILL, C.: Über Untersuchung des Geruchsinnes bei einzelnen Gruppen von Bediensteten der Wiener städtischen Gaswerke. Z. Österr. Vereines Gas- u. Wasserfachmännern **75**, 161—163 (1935).
LINNAEUS, C.: Odores medicamentorum. Amoenitates Academicae **3**, 183—201 (1756) (Defended by A. Wahlin).
LIPSITT, L. P., T. ENGEN, and H. KAYE: Developmental changes in the olfactory threshold of the neonate. Child Develop. **34**, 371—376 (1963).
LISZT, E. v.: Über den Geschmack- und Geruchsinn des Haussperlings. Ornithol. Beobachter **14**, 213—216 (1917).
LOCKE, B., and C. H. GRIMM: Odor selection, preferences and identification. J. appl. Psychol. **33**, 167—174 (1949).
LÖHNER, L.: Über menschliche Individual- und Regionalgerüche. Arch. ges. Physiol. 202, 25—45 (1924).
— Untersuchungen über die geruchsphysiologische Leistungsfähigkeit von Polizeihunden. Arch. ges. Physiol. **212**, 84—94 (1926).
— Von entscheidenden Versuchen über das Anzeigen von Witterungsübereinstimmung. Dtsch. Schäferhund **5**, 497—499 (1926).
— Zur Methodik geruchsphysiologischer Untersuchungen von Polizeihunden. Hund 1926, No. 10, 330—332.
— Über Fremdidentifizierung und Anzeigen von Witterungsübereinstimmung. Ztg. Vereins dtsch. Schäferhunde (SV) Sitz München **26**, 566—572 (1927).
— Grundsätzliches aus der Geruchsphysiologie. Ztg. Vereins dtsch. Schäferhund **30**, 422—436 (1931).
LOMBROSO, C., and M. CARRARA: Contributo all'antropologia dei Dinka. Atti Soc. Romana Antropol. **4**, 103—126 (1896/97).
—, u. S. OTTOLENGHI: Die Sinne der Verbrecher. Ztschr. Psychol. Physiol. Sinnesorgane **2**, 337—360 (1891).

LORRY: Observations sur les parties volatiles et odorantes des médicamens tirés des substances végétales et animales. Hist. Mém. Soc. Roy. Méd. **7**, 306—318 (1784/85).
LOWE, C. B.: Odor as an aid to the recognition of drugs, Amer. J. Pharm. **71**, 416—422 (1899).
LUCAS, G.: Premiers résultats d'une étude sur les produits odorants des calcaires fétides. C. R. Acad. Sci. (Paris) **234**, 121—123 (1952).
LUCHSINGER, R.: Der objective Nachweis des Geruchsvermögens (Olfacto-Pupillarreflex). Pract. oto-rhino-laryngol. **7**, 155—165 (1945).
LUTZ, K.: Beiträge zur Psychologie, Abrichtung und Verwendung des Diensthundes. Fortschr. Psychol. **5**, 257—291 (1920).
LUVARÀ, A., e M. MAURIZI: Ricerche di olfattometria in gravidanza. Boll. Mal. Orecch. **79**, 367—375 (1961).
MACLEOD, P.: Physiologie de l'olfaction et discrimination des odeurs. Concours Med. **87**, 1384, 1387, 1389, 1390, 1393, 1394 (1965).
— Variations de l'électro-olfactogramme et du potentiel lent glomérulaire en fonction de la nature du stimulus. Rev. Laryng. (Bordeaux) **86**, 855—859 (1965).
—, et A. CAVOY: Sensibilité de l'osmopile aux différents alcools primaires de la série aliphatique. J. Physiol. **52**, 158—159 (1960).
MAGENDIE, F.: Le nerf olfactif est-il l'organe de l'odorat? J. Physiol. exp. Pathol. **4**, 169—176 (1824).
MAHNER, P.: Vergleichende psycho-physiologische Versuche über die Unterscheidungsfähigkeit im Gebiete des inneren und äußeren Tastsinnes, des Geschmacks- und Geruchssinnes an taubstummen, blinden, normalsinnigen, schwachsinnigen und taubstumm-blinden Kindern. Bern: Dissertation 1908.
MALJEAN: L'Odeur de sainteté. Chron. Méd. **15**, 263 (1908).
MALVEZIN, P.: La composition chimique et le bouquet des vins. Ann. Chim. anal. Chim. appl. [ii], **4**, 298—301 (1922).
MARPLES, G.: Experimental studies on the ringed plover. Brit. Birds **25**, 34—44 (1931/32).
MARTIN-CALDERIN, A.: Nuestros estudios experimentales sobre la fisiologia de la olfaction. Siglo Med. **69**, 365—367, 395—397 (1922).
MASAAKI-ITO: Considérations expérimentales sur les fixations des odeurs. Rev. Laryng. (Bordeaux) **86**, 995 (1965).
MASON, A. H.: Odours, perfumes and flavours: physiology, philosophy, history, sources and preparation. Proc. Lit. Philos. Soc. Liverpool **37**, 195—217 (1882/83).
MASSERMAN, J. H., and C. PECHTEL: Normal and neurotic behavior in monkeys. A motion picture. J. nerv. ment. Dis. **124**, 518—519 (1956).
MATESON, J. F.: Olfactometry: its techniques and apparatus. J. Air Pollution Control Ass. **5**, 167—170 (1955).
MATHIEU, G.: Le chien criminel. Bruxelles: De Boeck 1911.
MATIGNON, J. J.: L'Odeur des anglais. Chron. Méd. **15**, 605 (1908).
MATTEOTTI, L.: Scala naturale degli odori. Boll. Mal. Orecch. **25**, 49—62 (1907). *Reprinted in* Riv. Ital. Ess. Profumi Piante Officin. Olii Veg. Saponi **28**, 88—93 (1946), *with notes by* Rovesti, 93—95.
MATTHES, E.: Die Leistungsfähigkeit des Geruchssinnes makrosmatischer Säugetiere. Forsch. Fortschr. **8**, 434—435 (1932).
— Geruchsdressuren an Meerschweinchen. Z. vgl. Physiol. **16**, 766—788 (1932).
— Weitere Geruchsdressuren an Meerschweinchen. Z. vgl. Physiol. **17**, 464—490 (1932).

— Olfacto e gôsto no reino animal. Arq. Museu Bocage (Lisboa) **9**, 17—46 (1938).
MATZKER, J.: Riechen und Lebensalter — Riechen und Rauchen. Umschau **65**, 673 (1965).
— Riechen und Lebensalter — Riechen und Rauchen. Arch. Ohren-, Nasen-, Kehlkopfheilk. **185**, 755—760 (1965).
MAY, J.: Geruchsschwellen von Lösemitteln zur Bewertung von Lösemittelgerüchen in der Luft. Staub **26**, 385—389 (1966).
McADAM, D. W., and J. S. WAY: Olfactory discrimination in the giant anteater. Nature **214**, 316—317 (1967).
McCARTNEY, W.: Quelques reflexions sur la litterature relative à l'olfaction. Parf. Cosm. Sav. **10**, 202—205 (1967).
McCORD, C. P., and W. N. WITHERIDGE: Odors: Physiology and Control. New York: McGraw-Hill 1949.
McDONALD, M. K.: An experimental study of Henning's system of olfactory qualities. Amer. J. Psychol. **33**, 535—553 (1922).
McKECHNIE, W. E.: Sense of smell in birds. Nature **110**, 784 (1922).
— Soaring flight and the "olfactory" organs of birds. Nature **111**, 48—49 (1923).
McKENZIE, D.: The riddle of olfaction. J. Laryngol. Otol. **36**, 288—296 (1921).
— Aromatics and the soul. A study of smells. London: Heinemann 1923.
McWHORTER, J.: The bloodhound as a witness. Amer. Law Rev. **54**, 109—122 (1920).
MEEK, A.: Sense of smell in birds. Nature **110**, 279 (1922).
MENZEL, R.: Die Arbeit mit Witterungskonserven. Z. Hundeforsch. Heft 1/2, 101—113 (1933).
MENZEL, RUDOLF, u. RUDOLFINE MENZEL: Das Problem der Verwendung des Hundes als „Feinriecher". Dtsch. Polizeihund **28**, 387—395 (1928).
— — Das „Mikrolfak" Hund und seine Leistungsfähigkeit. Ztg. Ver. dtsch. Schäferhunde **27**, 155—161 (1928).
— — Beobachtungen über das Abstraktionsvermögen des Hundes. Arch. ges. Psychol. **71**, 339—350 (1929).
— — Die Bedeutung der Gesetze über Schwellenwert und Reizsummation bei der Spürarbeit des Hundes. J. Psychol. Neurol. **38**, 258—264 (1929).
— — Der Eigengeruch des Hundes als Fehlerquelle bei Witterungsreinheits-Arbeiten. Ztg. Ver. dtsch. Schäferhunde (1929).
— — Die Verwertung der Riechfähigkeit des Hundes im Dienste der Menschheit. Berlin: Kameradschaft Verlagsgesellschaft 1930.
— — Tierpsychologische Voraussetzungen für die Nasenarbeit des Hundes. Z. Hundeforsch. **2**, 17—28 (1932).
— — Die Vererbung der Nasenleistungen und die sich daraus ergebenden Zuchtziele. Ber. Dritter kynologischer Weltkongreß 88—90 (1935).
MICHELS, K. M., D. S. PHILLIPS, R. H. WRIGHT, and J. PUSTEK: Odor and olfaction: a bibliography 1948—1960. Percept. Motor Skills **15**, 475—529 (1962).
MICHELSEN, W. J.: Procedure for studying olfactory discrimination in pigeons. Science **130**, 630—631 (1959).
MIDDLETON, A. W.: Smell — the physical sense. J. Soc. Cosm. Chem. **7**, 46—72 (1956).
— Smell — threshold concentration. J. Soc. Cosm. Chem. **8**, 41—43 (1957).
MILES, W. R., and L. H. BECK: Infrared absorption in field studies of olfaction in honeybees. Proc. nat. Acad. Sci. **35**, 292—310 (1949).
MILINE, R., V. DECEVERSKI et R. KRSTIC: Influence d'excitations olfactives sur le système habénulo-épiphysaire. Ann. endocrinol. **24**, 377—379 (1963).

Millen, J. K.: Your nose knows. A study of the sense of smell. Los Angeles: Cunningham Press 1960.
Miller, L.: Some tagging experiments with black-footed albatrosses. Condor **44**, 3—9 (1942).
Mody, N. H. N.: A collection of japanese clocks. London: Kegan Paul 1932 and Kôbe: J. L. Thompson 1932.
Moll, A.: Psychologische Untersuchungen an Spürhunden der Berliner Kriminalpolizei. Kriminalist. Monatsh. Suppl. Kriminalist. Probleme 2—16 (1927).
Moncrieff, A.: Biochemistry of mental defect. Lancet **II**, 273—278 (1960).
— Smell of "acetone". Lancet **II**, 552 (1960).
Moncrieff, R. W.: The Chemical Senses. London: Leonard Hill 1944; 2nd Ed. 1951. 3nd Ed. 1967.
— What is odor? A new theory. Amer. Perf. Ess. Oil Rev. **54**, 453—454 (1949).
— Loss of weight of odorants. Perf. Ess. Oil Rec. **44**, 346—347 (1953).
— The sorptive properties of the olfactory membrane. J. Physiol. **130**, 543—558 (1955).
— Olfactory adaptation and odour likeness. J. Physiol. **133**, 301—316 (1956).
— The sorptive nature of the olfactory stimulus. Proc. II. Internat. Cong. Surface Activity **2**, 160—167 (1957).
— An instrument for measuring and classifying odors. J. appl. Physiol. **16**, 742—749 (1961).
— Psycho-galvanic reflexes to odours. Perf. Ess. Oil Rec. **54**, 313—316 (1963).
— Stereochemical theory of odour. Soap, Perfumery, Cosmetics **38**, 1098, 1100 (1965).
— Changements, avec l'age des préférences olfactives. France Parf. **8**, 397—404 (1965).
— Odour Preferences. London: Leonard Hill 1966.
— Changes in olfactory preferences with age. Rev. Laryng. (Bordeaux) **86**, 895—904 (1965).
Monéry, A.: L'Ame des Parfums. Paris: Aristide Quillet 1924.
Montaigne: Michel E. de Montaigne. Essais de Messire Michel, Seigneur de Montaigne. Bordeaux: Millanges 1580.
de Montaigne, M. E.: Of smells. Essays. Trechmann's translation, London: Oxford Univ. Press 1927.
Morimura, S.: Untersuchung über den Geruchssinn. Tohoku J. exp. Med. **22**, 417—448 (1934).
Morris, R. C.: The sense of smell in Indian *Felidae*. J. Bombay nat. Hist. Soc. **33**, 695 (1929).
Most, K.: Neue Versuche über Spürfähigkeit. Hund Nos 18, 20, 22 (1926).
— Das Problem der Spurenreinheit auf der menschlichen Spur im Lichte der zumal mit der Fährtenbahn erzielten Versuchsergebnisse. Hund No. 2, 31—35 (1928).
— Die Abrichtung des Hundes. Braunschweig: Gersbach 1951.
— Über Fährtenarbeit des Hundes. Pirsch **4**, 120—124, 158—160 (1952).
— Training Dogs. London: Popular Dogs Pub. Co. 1954.
Most, K., u. G. H. Brückner: Die Vererbung der Nasenleistungen und die sich daraus ergebenden Zuchtziele. General-Ber. Dritter kynologischer Weltkongress 48—58 (1935).
— Über Voraussetzungen und den derzeitigen Stand der Nasenleistungen von Hunden. Z. Hundeforsch. N. F. 1, Heft 5 9—30 (1936).
Moulton, D. G.: Studies in olfaction. Dissertation. Univ. Birmingham. 1960.
— Studies in olfactory acuity. V. The comparative olfactory sensitivity of pigmented and albino rats. Animal Behav. **8**, 129—133 (1960).

MOULTON, D. G.: Studies in olfactory acuity. III. Relative detectability of *n*-aliphatic acetates by the rat. Quart. J. exp. Psychol. **12**, 203—213 (1960).
— Pigment and olfactory mechanism. Nature **195**, 1312—1313 (1962).
— and D. TUCKER: Electrophysiology of the olfactory system. Ann. N. Y. Acad. Sci. **116**, 380—428 (1964).
— Differential sensitivity to odors. Cold Spr. Harb. Symp. quant. Biol. **30**, 201—206 (1965).
— Physiological aspects of olfaction. J. Food Sci. **30**, 908—916 (1965).
—, and L. M. BEIDLER: Structure and function in the peripheral olfactory system. Physiol. Rev. **47**, 1—52 (1967).
— E. H. ASHTON, and J. T. EAYRS: Studies in olfactory acuity. IV. Relative detectability of *n*-aliphatic acids by the dog. Animal Behav. **8**, 117—128 (1960).
MOUQUET: Courtes observations sur l'osmologie ou histoire naturelle des odeurs de *M*. J.-J. Virey. Bull Pharm. **4**, 319—323 (1812).
MOZELL, M. M.: Electrophysiology of olfactory bulb. J. Neurophysiol. **21**, 183 (1958).
— Evidence for sorption as a mechanism of the olfactory analysis of vapours. Nature **203**, 1181—1182 (1964).
— Olfactory discrimination: electrophysiological spatiotemporal basis. Science **143**, 1336—1337 (1964).
MRAK, V., M. A. AMERINE, C. S. OUGH, and G. A. BAKER: Odor difference test with applications to consumer preferences. Food Res. **24**, 574—578 (1959).
MÜLLER, A.: Quantitative Untersuchungen am Riechepithel des Hundes. Z. Zellforsch. mikroskop. Anat. **41**, 335—350 (1954/55).
MUELLER, F.: Zum Problem der Schweißhund-Abrichtung. Z. Hundeforsch. **1**, Heft 3/4, 80—92 (1931).
— Die Verwendung des Jagdhundes nach dem Schuß. Z. Hundeforsch. N. F. **12**, Heft 4, 1—43 (1938).
MULLINS, L. J.: Olfactory thresholds for some homologous series of compounds. Fed. Proc. **14**, 105—106 (1955).
— Olfaction. Ann. N. Y. Acad. Sci. **62**, 247—276 (1955).
MURALT, v.: Die Gebrauchshundnase. Dtsch. Jäger-Ztg. **74**, 13—14 (1919/20).
MYERS, C. S.: Smell. I. Olfactory acuity and discrimination of odourstrengths. II. Memory and discrimination of odours. Rep. Cambridge Anthrop. Expedition to Torres Straits, **2**, part 2. Cambridge: Univ. Press 1903.
MYZNIKOV, N. M.: Čuvstviteljnostj obonjateljnogo analizatora služebnjich sobak i metody ee povysenija. — Zhurn. Vyss. Nerv. Deiat. Pavlova **8**, 744—750 (1958). English version in Pavlov. J. Higher Nerv. Activity **8**, 691—696 (1958).
NADER, J.: Symposium on odors. Current techniques of odor measurement. Arch. indust. Hlth. **17**, 537—541 (1958).
NAGEL, W.: Über Mischgerüche und die Komponentengliederung des Geruchssinnes. Z. Psychol. Physiol. Sinnesorgane **15**, 82—101 (1897).
NARODNY, L. H.: The resonance of smell. Per. Ess. Oil Rec. **47**, 23—28 (1956).
— The vector analysis of odours. Perf. Ess. Oil Rec. **48**, 31—32 (1957).
NAUŠ, A., and J. EDL: Jednoducha metoda ke zjistovani cichoveho prahu. Čs. Hyg. **6**, 618—623 (1961).
— and J. PIHRT: Nový zpusob olfaktometrie. Čs. Otolaryng. **5**, 290—295 (1956).
NAVES, Y.-R.: Odour and optical isomerism. Perf. Ess. Oil Rec. **42**, 40—41 (1951).
— La riproduzione sintetica della sostanze odorose naturali. Riv. Ital. Ess. Profumi Piante Officin. Olii Veg. Saponi **33**, 602—607 (1951).
— The relationship between the stereochemistry and odorous properties of organic substances. Soc. Chem. Indust. Monograph No. 1, pp. 38—51, Discussion 51—53. London: 1957.

NEGUS, V. E.: Observations on the comparative anatomy and physiology of olfaction. Acta oto-laryngol. **44**, 13—24 (1954).
— The air-conditioning mechanism of the nose. Brit. med. J. **I**, 367—371 (1956).
— The Comparative Anatomy and Physiology of the Nose and Paranasal Sinuses. London and Edinburgh: Livingston 1958.
NERGAARD, R.: Smak og lukt. T. Hermetikindust. **28**, 147—151 (1942).
NEUHAUS, W.: Die Bestimmung von Riechschwellen für Fettsäuren beim Hund. Proc. XIVth Internat. Cong. Zool. 296—298 (1956) 1953.
— Über die Riechschärfe des Hundes für Fettsäuren. Z. vergl. Physiol. **35**, 527 (1963).
— Die Riechfähigkeit bei Mensch und Tier. Umschau, **54**, 84—86 (1954).
— Die Riechschwellen des Hundes für Jonon und Äthylmercaptan und ihr Verhältnis zu anderen Riechschwellen bei Hund und Mensch. Z. Naturforsch. **9 b**, 560—567 (1954).
— Die Unterscheidung von Duftquantitäten bei Mensch und Hund nach Versuchen mit Buttersäure. Z. vgl. Physiol. **37**, 234—252 (1954/55).
— Die Riechschwelle von Mischdüften und die Spezifität der Riechreceptoren beim Hund. Verh. dtsch. zool. Ges. 66—71 (1955).
— Die Riechschwelle von Duftgemischen beim Hund und ihr Verhältnis zu den Schwellen unvermischter Duftstoffe. Z. vgl. Physiol. **38**, 238—258 (1955/56).
— Neuere Erfahrungen über die Riechfähigkeit des Hundes. Ztg. Ver. dtsch. Schäferhunde **50**, 197—206 (1956).
— Die Unterscheidungsfähigkeit des Hundes für Duftgemische Z. vergl. Physiol. **39**, 25—43 (1956).
— Wahrnehmungsschwelle und Erkennungsschwelle beim Riechen des Hundes im Vergleich zu den Riechwahrnehmungen des Menschen. Z. vergl. Physiol. **39**, 624—633 (1956/57).
— Unterschiede in der Riechschärfe bei Hunden. Z. vergl. Physiol. **40**, 65 (1957/58).
— Die Veränderlichkeit der Riechschärfe des Hundes durch orale Duftstoffgaben. Z. vergl. Physiol. **41**, 221—241 (1958).
— Der Eigengeruch des Menschen, seine Wahrnehmung, Bedeutung und Beeinflussung. Münch. med. Wschr. **103**, 1752—1755 (1961).
— u. D. M. GOLDENBERG: Über einige vegetative Reaktionen nach Geruchsreizen beim Menschen. Experientia **22**, 174—176 (1966).
— und A. MÜLLER: Das Verhältnis der Riechzellenzahl zur Riechschwelle beim Hund. Naturwiss. **41**, 237 (1954).
NEURATH, H.: Über die Leistung des Geruchssinnes bei Elritzen. Z. vergl. Physiol. **31**, 609—626 (1949).
NICCOLINI, P.: Ulteriori indagini sulla probabile importanza dell'ossigeno nella percezione degli odori. Boll. Soc. ital Biol. Sper. 8, 424—426 (1933).
— Schema di una suddivisione degli elementi chimici nei riguardi della loro olfattosensibilita. Boll. Soc. ital. Biol. sper. **9**, 369—370 (1934).
— Distribuzione degli elementi olfattopositivi in alcune classificazioni chimiche. Boll. Soc. ital. Biol. sper. **9**, 370—371 (1934).
— Relazioni fra stimoli semplicemente odorosi o odorosi-irritanti e concentrazione solfocianica del muco nasale. Boll. Soc. ital. Biol. sper. **10**, 429—430 (1935).
— Modificazioni della concentrazione solfocianica nel muco nasale ed in altre secrezioni, in relazione con stimolazioni, del trigemino e dell'olfattorio. Boll. Soc. ital. Biol. sper. **10**, 431—432 (1935).
— Contributo allo studio fisio-farmacologico dell'olfatto. II. Reazioni biologiche allo stimolo di sostanze olfattosensibili. (ii). Sul probabile meccanismo di azione del componente solfocianico del muco nasale in rapporto alla funzione del senso dell'olfatto. Arch. ital. sci. Farmacol. **7**, 297—310 (1938).

NICCOLINI, P.: Sguardo d'insieme sullo stato attuale delle mie ricerche sul senso dell'olfatto. Margin. Otolaryngol. **1,** 154—162 (1940).
— Problemi insoluti intorno all'olfatto: sospetto di relazioni fra sostanze olfattosensibili, pigmento dell'epitelio olfattorio e redoxasi nasali. Atti R. Accad. Fisiocrit. (Siena) [xi], **8,** 106—111 (1940).
— L'Olfatto. Ipotesi che tendono ad interpretare il meccanismo di percezione degli odori, e ricerche destinate ad avvalorarle. Arch. Psicol. Neurol. Psichiat. **3,** 31—78 (1942).
— La stimolo olfattorio e la sua recezione. Arch. ital. sci. Farmacol. [iii], **4,** 109—172 (1954).
— Modern theories of osmogenic-neural energy transformation as the initial phenomenological basis of the olfactory process. Ital. gen. Rev. Oto-Rhino-Laryngol. **2,** 218—278 (1960).
NICHOLS, F. L., and E. H. S. BAILEY: The sense of smell. Nature **35,** 74—75 (1886/87).
NICKLÈS, J.: Sur la théorie physique des odeurs et des saveurs. Mém. Soc. (Roy.) Sci. Belles Lettres Nancy **24,** 356—385 (1861).
NIELSEN, A.: Das Riechen als analytische Methode. Chem.-Ztg. **64,** 389—391 (1940).
NIEMAND, H. G.: Ist der Individualgeruch für den Hund richtunggebend zum Erkennen seines Herrn? Z. Hundeforsch., N. F. **12,** pp. 61 (1938).
NOGUÉ, J.: Essai d'une description du monde olfactif. J. Psychol. norm. Pathol. **33,** 230—275 (1936).
— Esquisse d'un Systeme des Qualités Sensibles. Paris: Presses Universitaires de France 1943.
NOLTE, W.: Zum Geruchsvermögen der Enten. Zool. Anzeiger **71,** 115—135 (1927).
OGLE, W.: Anosmia or cases illustrating the physiology and pathology of the sense of smell. Med. Chir. Trans. **53,** 263—290 (1870).
OHMA, S.: La classification des odeurs aromatiques en sous-classes. Arch. néerland. Physiol. **6,** 567—590 (1921/22).
OLDFIELD, C. A.: The phenomena of animal scent. Perf. Ess. Oil Rec. Catalogue 13—17 (1936).
OLIVIER, H. D.: Sense of smell in the lion. J. Bombay nat. Hist. Soc. **33,** 977 (1929).
ONAGAWA, K.: I. On the relationship between the grade of colour at the olfactory mucous membrane and the sensibility of olfactory stimulation. II. On the relationship between the action potential of the olfactory mucous membrane and the grade of colour of the olfactory mucous membrane. III. On the relationship between the frequency of the spike discharge from the bulbus olfactorius and the grade of colour of the olfactory mucous membrane. J. physiol. Soc. Japan **19,** 189—193, 194—197, 198—205 (1957).
OTTOLENGHI, S.: L'Olfatto nei criminali. Giorn. Accad. Med. Torino [iii], **36,** 427—436 (1888).
OTTOSON, D.: Studies on the relationship between olfactory stimulating effectiveness and physico-chemical properties of odorous compounds. Acta physiol. scand. **43,** 167—181 (1958).
— Some aspects of the function of the olfactory system. Pharmocol. Rev. **15,** 1—42 (1963).
— Mécanismes électro-physiologiques de transmission des messages olfactifs. Rev. Laryng. (Bordeaux) **86,** 832—844 (1965).
OUCHAKOV, A. A.: L'Adsorption des substances odorantes et l'odorat. Acta otolaryngol. **14,** 470—511 (1930).
— Essais d'application pratique concernant la relation entre l'odorat et l'absorption des substances odorantes. Rev. Laryngol. Otol. Rhinol. **51,** 77—91 (1930).

OWRE, O. T., and P. O. NORTHINGTON: Indication of the sense of smell in the turkey vulture, *Cathartes aura* (Linnaeus), from feeding tests. Amer. Midland Naturalist **66**, 200—205 (1961).

PALLESTRINI, E.: Nuovi orientamenti fisico-elettrici in olfattometria. Min. Otorinolaringol. **1**, 119—135 (1951).

PALMER, C. W.: Powers of perception in birds. Nature **112**, 688 (1923).

PARKER, G. H.: Olfactory reactions in fishes. J. exp. Zool. **8**, 535—542 (1910).

— The olfactory sense of fishes. Amer. J. Physiol. **27**, xix (1910/11).

— The olfactory reactions of the common killifish *Fundulus heteroclitus* (Linn.) J. exp. Zool. **10**, 1—5 (1911).

— The directive influence of the sense of smell in the dogfish. Bull. U.S. Bur. Fish. **33**, 61—68 (1914/15).

— Smell, taste, and allied senses in the vertebrates. Philadelphia and London: Lippincott n. d.

— and R. E. SHELDON: The sense of smell in fishes. Bull. U.S. Bur. Fish. **32**, 33—46 (1912/14).

— and E. M. STABLER: On certain distinctions between taste and smell. Amer. J. Physiol. **32**, 230—240 (1913).

PARKES, A. S.: Olfactory and gustatory discrimination in man and animals. Proc roy. Soc. Med. **56**, 47—51 (1963).

—, and H. M. BRUCE: Olfactory stimuli in mamalian reproduction, Science **134**, 1049—1054 (1961).

PARRY, E. J.: The classification of odours. Perf. Ess. Oil Rec. **7**, 129—132 (1916).

PASSY, J.: Sur les minimums perceptibles de quelques odeurs. C. R. Acad. Sci. (Paris) **114**, 306—308 (1892).

— Sur quelques minimums perceptibles d'odeurs. C. R. Acad. Sci. (Paris) **114**, 786—788 (1892).

— Revue générale sur les sensations olfactives. Ann. psychol. **2**, 363—410 (1895).

PASTIER, J.: Contribution à l'etude de la psychologie de l'olfaction. Limoges: Bontemps 1939.

PATTERSON, D. S., P. TURNER, and J. V. SMART: Smell threshold in Diabetes mellitus. Nature **209**, 625 (1966).

PATTERSON, P. M., and B. A. LAUDER: The incidence and probable inheritance of "smell blindness" to normal butyl mercaptan. J. Hered. **39**, 295—297 (1948).

PAULING, L.: Analogies between antibodies and simpler chemical substances. Chem. Eng. News **24**, 1064—1065 (1946).

PECHTEL, C., and J. H. MASSERMANN: The osmatic responses of normal and neurotic monkeys. Ann. N. Y. Acad. Sci. **58**, 256—260 (1954).

PENROSE, F. G.: Field notes on a pair of stone-curlews. Have they an appreciable sense of smell? Brit. Birds **6**, 266—269 (1912/13).

PEREZ, B.: Les Trois Premières Années de l'Enfant. Paris: Alcan. 1 Éd. 1878, 5 Éd. 1892.

PETERSON, F., and L. H. RAINEY: The beginnings of mind in the newborn. Bull. Lying-In Hosp. City N. Y. **7**, 99—122 (1910/11).

PEYRON, L.: Edifices moléculaires et odeurs. France Parfums **7**, 397—405 (1964).

PFAFFMANN, C., and T. ENGEN: Symposium on odors. Psychology: absolute judgments of odor. Arch. indust. Hlth **17**, 544—545 (1958).

PFUNGST, O.: Neue Dressurergebnisse. J. Psychol. Neurol. **26**, 318—320 (1920/21).

PHILIPPOT, E., et M. A. GEREBTZOFF: Premiers resultats de l'analyse du pigment olfactiv. J. Physiol. **50**, 451—452 (1958).

PICCARD, J., et A. PICCARD: Phénomènes odorifiques produits par le choc. Arch. Sci. Phys. Nat. [iv], **25**, 425—429 (1908).

Pick, W. H.: A note on the meteorological conditions affecting scent. Meteorological Mag. **57**, 16—17 (1922).

Piesse, C. H.: Olfactics and the physical senses. London: Piesse & Lubin 1887.

Piesse, G. W. S.: The Art of Perfumery. London: Longmans 1855.

Pipping, M.: Der Geruchssinn der Fische mit besonderer Berücksichtigung seiner Bedeutung für das Aufsuchen des Futters. Comment. Biol. Soc. Sci. Fennica **2**, No. 4, 28 pp., No. 10, 10 pp. (1924/27).

Plank, R.: Der gegenwärtige Stand der Klassifikation und objektiven Bewertung von Geschmacks- und Geruchsempfindungen. Z. Ver. dtsch. Chemiker, Beiheft No. 52, 19 pp. (1945).

Pliška, V., u. G. Jáníček: Olfactometrische Studien. I. Die Definition der Wahrnehmungsschwelle-Konzentration mittels Volumengraden. Sborn. Vys. Skol. Chem.-Techn. Praze Potrav. Technol. **4**, 115—127 (1960).

— — Die Veränderungen der Wahrnehmungs-Schwellen-Konzentration der Riechstoffe in einigen homologischen Serien. Arch. int. Pharmacodyn. **156**, 211—216 (1965).

— P. Dusek, u. G. Jáníček: Olfaktometrische Studien. II. Die Wahrnehmungsschwellen-Konzentration des Gemisches zweier Riechstoffe. Sborn. Vys. Skol. Chem.-Techn. Praze Potrav. Technol. **5**, 11—23 (1960).

Pocock, R. I.: Taste or smell in the laughing jackass (Dacelo). Nature **89**, 425 (1912).

— Scent-glands in mammals. Proc. Zool. Soc. London, 742—755 (1916).

Pollard, H. B. C.: The mystery of scent. London: Eyre & Spottiswoode 1937.

Posvic, H.: The odors of optical isomers. Science **118**, 358—359 (1953).

Poucher, W. A.: A classification of odours and its uses. J. Soc. Cosmetic Chemists **6**, 80—94 (1955).

Pradines, M.: Sur l'objectivité des odeurs et des saveurs. J. Psychol. norm. Pathol. **26**, 16—73 (1929).

Pressler, C. E.: Theories on odors. Drug Cosmetic Indust. **62**, 180—182, 273—274 (1948).

Prevost, B.: Sur les émanations des corps odorans. Ann. Chim. **21**, 254—261 (1797); **22**, 111—112 (1797).

— Sur les effets odoroscopiques des corps chauds et des corps froids inodores, et sur les substances odorantes. Ann. Chim. **24**, 31—56 (1797).

Prevost, J. L.: Atrophie des nerfs olfactifs, fréquente chez le vieillard, et correspondant avec la diminution ou la perte du sens de l'odorat. Gaz. méd. (Paris) [iii], **21**, 597—600 (1886).

Preyer, W.: Die Seele des Kindes. 5. Aufl. Leipzig: Griebens Verlag 1900.

Primavori, G. G.: Il muco nasale come probabile attualizzatore dell'odore potenziale di alcuni composti. Bull. Soc. ital. Biol. sper. **30**, 266—267 (1954).

Prince. R. G. H., and J. H. Ince: The measurement of intensity of odour. J. appl. Chem. **8**, 314—321 (1958).

Proceedings of the NATO Advanced Study Institute on Odor and Theories of Odor Measurement: Theories of Odor and Odor Measurement. Istanbul 1966: N. Janyolac 1968.

Pumphrey, R. J.: The sense organs of birds. Ibis **90**, 171—199 (1948).

Pustet, A., u. K. Berger: Versuche über das Gedächtnis und das Lernvermögen der Tauben. Z. vergl. Physiol. **9**, 668—684 (1929).

Quix, F. H.: Olfaktometrische Untersuchungsmethoden. Internat. Centralbl. Ohrenheilk. **3**, 57—71 (1905).

Ramsay, W.: On smell. Proc. Bristol Naturalists' Soc. **3**, 280—293 (1879/82); Nature **26**, 187—189 (1882).

RANDEBROCK, R.: Das Polaritätsprofil als Mittel zur Geruchsbeurteilung. J. Soc. Cosmetic Chemists **16**, 653—677 (1965).
RASPAIL, X.: Le sens de l'odorat chez les oiseaux. Rev. Sci. [iv] **12**, 144—148 (1899). English translation: Ann. Rep. Smithsonian Inst. 367—373 (1899, 1901).
RAUB, E.: Wenig bekannte Eigenschaften der Metalle. Mitt. Forsch. Inst. Probieramts Edelmet. Staatl. höheren Fachschul, Schwäb.-Gmünd. **7**, 51—64 (1933/34).
— Die Entstehung des Metallgeruchs und -geschmacks. Z. angew. Chem. **47**, 673—675 (1934).
RAWDON-SMITH, A. F.: Theories of Sensation. Cambridge: Univ. Press 1938.
RAZRAN, H. S., and C. J. WARDEN: The sensory capacities of the dog as studied by the conditioned reflex method (Russian Schools) Psychol. Bull. **26**, 202—222 (1929).
REEKER, H.: Zum Geruchsinn der Vögel. Jahresber. westf. Prov.-Ver. Wiss. Kunst **27**, 44—45 (1899).
REESE, T. S., and S. S. STEVENS: Subjective intensity of coffee odor. Amer. J. Psychol. **73**, 424—428 (1960).
RÉGISMANSET, C.: Philosophie des Parfums. Paris: Sansot 1907.
REID, T.: An Enquiry into the Human Mind on the Principles of Common Sense. Edinburgh and London: Neill & Co. 1801.
REISER, O.: Wie finden die Geier das Aas? Verh. V. Internat. Ornithol. Kongr. 561—562 (1910).
REISS: Der Spürhund als Hilfsmittel in der Kette der Ermittlungsmethoden. Kriminal. Monatsh. **3**, 250—252 (1929).
REPOND, A.: Odeur de sainteté et tumeur cérébral. Rev. méd. Suisse Romande **46**, 49—52 (1926).
REŞAT ALI: Koku alimi duygusu. Askeri Sihhiye Mecmuasi **64**, 24—36 (1935).
RHOADS, S. N.: The power of scent in the turkey vulture. Amer. Naturalist **17**, 829—833 (1883).
RICHARDS, G. G.: Are we neglecting the art of physical diagnosis in our modern medical education? Ann. int. Med. **22**, 362—367 (1942).
RICHARDSON, E. H.: The homing instinct in dogs. Psyche **2**, 52—56 (1921/22).
— War, Police and Watch Dogs. London: Blackwood 1910.
— Forty Years with Dogs. London: Hutchinson n. d.
RIMMEL, E.: The Book of Perfumes. 2nd Ed. London: Chapman & Hall 1865.
RIVERS, W. H. R.: Observations on the senses of the Todas. Brit. J. Psychol. **1**, 321—396 (1904/05).
ROBERTSON, G. C.: External perception in dogs. Nature **7**, 322—323 (1872/73).
ROBERTSON, P. W.: Do smells have frequencies? New Scientist **11**, 419 (1961).
ROBINSON, D. Z.: Theory of smell. New Scientist **5**, 714 (1959).
ROMANES, G. J.: Experiments on the sense of smell in dogs. Nature, **36**, 273—274 1887).
ROSANO, H. L., and S. Q. SCHEPS: Adsorption-induced electrode potential in relation to olfaction. Ann. N. Y. Acad. Sci., **116**, 590—601 (1964).
— Objective approaches to odor measurement. Adv. Chem. Ser. **56**, 53—63 (1966).
ROSENBERG, B., T. N. MISRA, and R. SWITZER: Mechanism of olfactory transduction. Nature **217**, 423—427 (1968).
ROSS, S., and A. E. HARRIMAN: A preliminary study of the Crocker-Henderson odor-classification system. Amer. J. Psychol. **62**, 399—404 (1949).
ROUDNITSKA, E.: Le parfum dans le système des beaux-arts. Indust. Parf. **8**, 5 (1953).
— La perceptibilité des odeurs. Parfumerie **2**, 68—73 (1944).
ROUGET, J. M. P. J.: Les possibilités actuelles de l'olfactométrie clinique. Rev. Laryngol. Otol. Rhinol. **82**, 543—619 (1961).

Rous, J.: Elektrofyziologie čichového analyzátoru. Čs. Otolaryngol. **13**, 179—189 (1964).
—, and V. Synek: Objektivni stanoveni intenzity čichového vjemu pomoci galvanického kožního reflexu. Čs. otolaryngol. **15**, 271—278 (1966).
Rouse, J. E.: Respiration and emotion in pigeons. J. comp. Neurol. Psychol. **15**, 494—513 (1905).
Rovesti, P.: Some psycho-physiological and social aspects of the use of perfumes. Soap, Perfumery, Cosmetics **38**, 143—144, 136 (1965).
Royce, J. R.: On the mental ability of the dog. Science **110**, 666 (1949).
Rubin, M., D. Apotheker, and R. Lutmer: The stereochemical theory of olfaction. III. Structure and odor: 1,4-cyclohexane lactones and related compounds. Proc. Sci. Sect. Toilet Goods Ass. No. 37, Suppl. 24—33 (1962).
Rudolph, K.: Über den „Geruchsinn" der Hunde. Psyche **7**, 99—100 (1922/23).
Rüdel, O.: Ein sicheres Vorzeichen des Todes. (Münch. med. Wschr. **66**, 1422 (1919).
Rupe, H., u. K. von Majewski: Über osmophore Gruppen. Ber. dtsch. chem. Ges. **33**, 3401—3410 (1900).
Russell, T.: Egyptian Service 1902—1946. London: Murray 1949.
Russell, W. J.: The sense of smell in dogs. Nature **36**, 317 (1887).
Rutherford Food Corporation: Objective method of testing the pungency and odor of organic substances. U.S. Patent, May 24, No. 3, 252, 873 (1966).
Ruzicka, L.: Über die Henningschen Geruchsklassen. Dtsch. Parf.-Ztg. **5**, 173—175 (1919).
— Die Grundlagen der Geruchschemie. Chemiker-Ztg. **44**, 93—94; 129—131 (1920).
— Fundamentals of odour chemistry: a summary. Soc. Chem. Indust., Monograph No. 1, pp. 116—124. London: 1957.
Sacher, J. F.: Maßanalytische Bestimmungen mit Hilfe des Geruchsinnes. Chem.-Ztg. **37**, 1222 (1913).
Sagarin, E.: The Science and Art of Perfumery. New York: McGraw-Hill 1945.
— On the inherent invalidity of all current systems of odor classification. J. Soc. Cosmetic Chemists **2**, 25—35 (1950).
— Toward scientific method in the study of olfaction. Proc. Sci. Sect., Toilet Goods Ass. No. 14 (1950).
Sahlstedt, A. V.: Beitrag zur Kenntnis des Geruchsmechanismus bei den makrosmatischen Säugetieren. Skand. Arch. Physiol. **28**, 1—12 (1913).
Saint-Hilaire, G.: Mémoire sur la structure et les usages de l'appareil olfactif dans les poissons, suivi de considérations sur l'olfaction des animaux qui odorent dans l'air. Arch. Sci. Nat. **6**, 322—354 (1825).
Sakharoff, N. A.: O khimicheskom processe, lezascem v osnove deistviya organa obonianija. Protokol. Zasaid. Imp. Kavkazsk. Med. Obsh. **37**, 496—499. Discussion 500—502 (1900/01).
— O khimize deistviya organa obonianija i nervnikh processov voobsce. Protokol. Zasaid. Imp. Kavkazsk. Med. Obsh. **38**, 186—191; 235—248 (1901/02).
Salverda-Ter Laag, P. B.: L'Eau que prefèrent les oiseaux pour boire et pour se baigner. Arch. néerl. Physiol. **21**, 294—308 (1936).
Salwey, C. M.: On pastimes and amusments of the Japanese. Trans. Proc. Jap. Soc. (London) **5**, 76—93 (1898/1901).
de Sanctis, S.: Applicazione della legge di Weber-Fechner all'olfatto. Arch. ital. Biol. **36**, 131—132 (1901/02).
Sansot, E.: Essai sur les Parfums. Paris: Chiberre n. d. Nice: Modern-Studio n. d.
Sarris, E. G.: Hunde ausbilden — nicht abrichten. Stuttgart: Franckh'sche Verlagshandlung 1938.

SASS, O. E.: Über die Riechstoffe zu den Lipoiden. Parf. Kosmet. **33**, 400—402; 442—444 (1952).
SAUNDERS, H. C.: The stereochemical theory of olfaction. V. Some odor observations in terms of the Amoore theory. Proc. Sci. Sect., Toilet Goods Ass. No. 37, Suppl. 46—47 (1962).
SAVELIEV, N. A.: Fisiologiya nervi olfactorii Istoricheskiya i Eksperimentalniya Izsliedovaniya. Moskva: Univ. Tipograf. 1892.
— Fiziologiya obonianiya istoricheskiya i eksperimentalniya izsliedovaniya. Trud. Fiziol. Lab. Imp. Moskov. Univ. **4**, 121—378 (1893).
SAWER, J. C.: Rhodologia: a Discourse on Roses and the Odour of the Rose. Brighton: Smith 1894.
SAYLES, I.: The sense of smell in *Cathartes aura.* Auk N. S. **4**, 51—56 (1887).
SCHAFFER, J.: Rehfährte und Hund. Z. mikr.-anat. Forsch. **43**, 15—16 (1938).
SCHEIDT, W.: Anthropologie. 40—41. Die sogenannten niederen Sinne. Hamburg: Anthropologisches Inst. Univ. 1953.
SCHELENZ, H.: „Düfte in Mineralreich?" Dtsch. Parf.-Ztg. **6**, 211—212 (1920).
SCHILLER, P. VON: Intersensorielle Transposition bei Fischen. Z. vgl. Physiol. **19**, 304—309 (1933).
SCHLEMMER: Gebrauchshundnase. Dtsch. Jäger-Ztg. **74**, 475 (1919/20).
SCHLEY, O.-H.: Untersuchung über Geruchsschwellen. Würzburg: Dissertation 1934.
SCHMALFUSS, H.: Über Geruch und Riechen als Schlüssel zum Verständnis der Geruchswelt des Hundes. Terrier **47**, 80 (1954).
SCHMID, B.: Wie weit sieht der Hund, und auf welche Entfernung erkennt er seinen Herrn? Z. Hundeforsch. **3**, 188—208 (1933).
— Nasenleistung und psychische Fähigkeiten des Hundes. Umschau **39**, 457—460 (1935).
— Über die Ermittelung des menschlichen und tierischen Individualgeruches durch den Hund. Z. vgl. Physiol. **22**, 524—538 (1935).
— Carnivoren-Studien. I. Zur Psychologie der Caniden. Wolf — Hund — Fuchs. Kleintier Pelztier **12**, Heft 6, 77 pp. (1936).
— Hunde ermitteln menschlichen und tierischen Eigengeruch. Tierärztl. Rundschau **42**, 689—692 (1936).
— Umfang und Grenzen der Nasenleistung von Hunden. Forsch. Fortschritte **13**, 26—27 (1937).
— Nachweis des menschlichen und hundlichen Individualgeruches durch den Hund. Ztg. Fachschaft dtsch. Schäferhunde **36**, 58—63 (1937).
— Zur Psychologie unseres Hundes und seine Verwendung im Kriege. Umschau **44**, 516—521 (1940).
SCHMIDT, F.: Verbrecherspur und Polizeihund. Praktische Versuche, Beobachtungen und Winke für Polizeihundmänner u. Behörden. Augsburg: Selbstverlag des Vereins für deutsche Schäferhunde 1910.
— Die menschliche Geruchspur und die Fähigkeiten des Polizeihundes. Jber. naturwiss. Vereins Osnabrück (1907/1910) **17**, 31—55 (1911).
SCHMIDT, H.: Zur Frage der Entstehung der Geruchsempfindungen. Prometheus **25**, 749 (1913/14).
SCHMIDT, J.: On the aroma in plants raised by crossing. C. R. Trav. Lab. Carlsberg **11**, 330—332 (1914/17).
SCHNEIDER, R. A.: The relation of nasal membrane function to olfactory acuity. Fed. Proc. **19**, 301 (1960).
— J. P. COSTILOE, A. VEGA, and S. WOLF: Olfactory threshold technique with nitrogen dilution of *n*-butane and gas chromatography. J. appl. Physiol. **18**, 414—417 (1963).

SCHNEIDER, R. A., R. P. HOWARD, and S. WOLF: Olfactory perception thresholds in hypogonadal women: changes accompanying administration of androgen and estrogen. J. clin. endocrinol. Metab. **18**, 379—390 (1958).

—, C. E. SCHMIDT, and J. P. COSTILOE: Relation of odor flow rate and duration to stimulus intensity needed for perception. J. appl. Physiol. **21**, 10—14 (1966).

—, and S. WOLF: Olfactory perception thresholds for citral utilizing a new type olfactorium. J. appl. Physiol. **8**, 337—342 (1955).

— Relation of olfactory acuity to nasal membrane function J. appl. Physiol. **15**, 914—920 (1960).

— — B. DOLENZ, and J. WILLIAMS: Odor perception thresholds for "citral" utilizing a new type olfactorium. J. Lab. clin. Med. **44**, 927—928 (1954).

SCHNITZER, S. B., M. E. SCHNITZER, and J. L. FULLER: Studies on the tracking behavior of dogs. II. Breed differences in locating a hidden animal. Chicago: Armour Res. Inst. 1957.

SCHUMACHER, S.: Das Stirnorgan des Rehbockes *(Capreolus capreolus,* L.), ein bisher unbekanntes Duftorgan. Z. mikroskop.-anat. Forsch. **39**, 215—230 (1936).

SCHUTZ, H. G.: A matching-standards method for characterizing odor qualities. Ann. N. Y. Acad. Sci. **116**, 517—526 (1964).

— L. I. HILLENBRAND, and H. W. RAY: Relations of odor to physico-chemical properties of compounds. ASTIA Document 237 834, Contract No. DA 19-129-qm-1500.

SCHWARTZ, F.: Über das Auftreten des Geruches beim Reiben von Mineralien. Zbl. Mineral. Geol. Paläontol. 660—665 (1913).

SCHWARTZ, M.: Beiträge zur Ernährungsbiologie unserer körnerfressenden Singvögel. Arb. kaiserl. biol. Anst. Land-Forstwirtsch. **6**, 445—486 (1908).

SCOTT-MONCRIEFF, R.: Olfaction. Bull. Vancouver med. Ass. **15**, 266—274 (1939).

SEEGER: Der Spürhund. Kriminal. Monatsh. **3**, 202—205 (1929).

SEFFRIN, L.: Über die kleinsten noch wahrnehmbaren Geruchsmengen einiger Riechstoffe beim Hund. Z. Biol. **65**, 493—512 (1914/15).

SEIFERLE, E.: Zur Psychologie des Hundes. Berl. tierärztl. Wschr. No. 11, 163—165 (1938).

SEMERIA, C.: La olfattometria per via ematogena. Contributo allo studio patogenetico e diagnostico delle iposmie. Arch. ital. Otol. Rinol. Laringol. **66**, 435—444 (1955).

— Studio delle reazioni psicogalvanometriche alla stimolazione olfattiva. Olfattometria obbiettiva. Min. Otorinolaringol. **6**, 97—101 (1956).

— Il riflesso psicogalvanico alla stimolazione olfattiva. Riv. Oto-Neuro-Oftalmol. **31**, 98—103 (1956).

— La sensazione olfattiva ematogena. Arch. ital. Otol. Rinol. Laringol. **68**, Suppl. 30, 78 pp. (1957).

SEM-JACOBSEN, C. W., R. G. BICKFORD, H. W. DODGE, and M. C. PETERSEN: Human olfactory responses recorded by depth electrography. Proc. Staff Meet. Mayo Clin. **28**, 166—170 (1953).

SEVERN WILDFOWL TRUST: The Scientific work of the Wildfowl Trust. April 1957.

SHARLAND, M.: Swamp-harrier's sense of smell. Emu **47**, 81—83 (1947/48).

SHEARER, W. M.: 'Homing instinct' in sea trout. Nature **176**, 171—172 (1955).

SHELDON, R. E.: The reactions of the dogfish to chemical stimuli. J. comp. Neurol. Psychol. **19**, 273—311 (1909).

— The sense of smell in selachians. J. exp. Zool. **10**, 51—62 (1911).

SHERMAN, A. R.: Experiments in feeding humming birds during seven summers. Ann. Rep. Smithsonian Inst. 459—468 (1914).

SHKAPENKO, G., et M. A. GEREBTZOFF: Critique expérimentale d l'intervention des radiations dans le mécanisme de l'olfaction. Arch. int. Physiol. **59**, 423—429 (1951).

SKRAMLIK, E. VON: Über das Verhalten des Geruchsinns bei gleichzeitiger Einwirkung zweier Reize. Klin. Wschr. **2**, 1250—1253 (1923).

— Die physiologische Charakteristik von riechenden Stoffen. Naturwiss. **12**, 813—824 (1924).

— Über die Lokalisation der Empfindungen bei den niederen Sinnen. Z. Sinnesphysiol. **56**, 69—140 (1925).

— Handbuch der Physiologie der niederen Sinne. I. Die Physiologie des Geruchs- und Geschmackssinnes. Leipzig: Thieme 1926.

— Eine Vorrichtung zur Prüfung der Leistungen des Geruchssinnes. Münch. med. Wschr. **83**, 1564—1567 (1936).

— Verfahren zur Prüfung der Leistungen des Geruchssinnes. Berlin und Wien: Urban und Schwarzenberg 1937. Abderhaldens Handbuch der biologischen Arbeitsmethoden, Abt. V, Teil 7, Heft 2.

— Über die zur minimalen Erregung des menschlichen Geruchs- bzw. Geschmackssinnes notwendigen Molekülmengen. Arch. ges. Physiol. **249**, 702—716 (1948).

— Über Geruchs- und Geschmacksstärke. Z. Biol. **113**, 227—240 (1962).

— Über die Zahl der Empfindungsqualitäten im Gebiete der chemischen Sinneswerkzeuge (Geruchs- und Geschmackssinn). Z. Biol. **113**, 329—339 (1963).

SMEKAL, A.: Über den im Hochgebirge bei Steinschlag auftretenden Geruch. Naturwiss. **8**, 496 (1920).

SMITH, C. G.: Age incidence of atrophy of olfactory nerves in man. A contribution to the study of the process of ageing. J. comp. Neurol. **77**, 589—595 (1942).

SMITH, G. E.: The significance of the cerebral cortex. Brit. med. J. **I**, 758, 796—797 (1919); **II**, 11—12 (1919).

SMOLÉN, K. Czym jest zapach. Wszechświat No. 7/8, 214—217 (1958).

— Wrażliwość czlowieka i zwierząt na zapachy. Wszechświat No. 9, 258—260 (1958).

Société du Gaz de Paris: Procédé et appareil pour la détermination quantitative et qualitative du pouvoir odorant des corps volatils. French Patent, No. 607 612 (1925).

Société Nationale pour l'Amélioration du chien de Berger Belge: Programme des Épreuves en Campagne pour Chiens de Berger Belges qui auront lieu les 13 et 14 Avril 1912. Bruxelles: La Société 1912.

SOUDEK, Š.: K otázce čichu u ptáků. Biol. Listy **11**, 1—23 (1925).

— The sense of smell in the birds. Rapports X Congr. Internat. Zool. **1**, 755—765 (1927).

SPARHAWK, C. V.: What's the truth about odour? Perf. Ess. Oil Rec. **36**, 187—190 (1945).

SRNETZ, A.: Beitrag zur Physiologie des Riechaktes. Wien. tierärztl. Mschr. **22**, 753—767 (1935).

STAGER, K. E.: The role of olfaction in food location by the turkey vulture *(Cathartes aura)*. Dissert. Absts. **25**, 2115 (1964).

— The role of olfaction in food location by the turkey vulture *(Cathartes aura)*. Los Angeles County Mus., Contrib. Sci., No. 81, 63 pp. (1964).

STALMANS, G. M. J.: Use of police dogs in Ghent. Private communication 1959, Sept. 21.

Stark, J.: On the influence of colours on odours. Phil. Trans. **123**, 300—312 (1833).
— Historical account of experiments regarding the influence of colour on heat, the deposition of dew, and odours. Edinb. New Philos. J. **17**, 65—98 (1834).
Stefanini, A.: Sulla teoria della percezione degli odori. Arch. ital. Otol. Rinol. Laringol. **33**, 144—154 (1922).
— Sur la théorie de la perception des odeurs. Arch. ital. Biol. **74**, 8—17 (1924).
Stein, V. S.: Rhinologiske Funktionsundersögelser. Hospitalstidende, **43**, 1068—1081, 1091—1101, 1131—1138 (1900).
Steiner, G.: Geruchssinn und Orientierung der Tiere. Universitas **12**, 73—80 (1957).
Stephanitz, M. von: The German Shepherd Dog. 8th revised and enlarged Ed. Augsburg: Verein für deutsche Schäferhunde (SV) 1950.
Stephenson, W.: The inverted factor technique. Brit. J. Psychol. **26**, 344—361 (1935/36).
Steven, D. M.: Studies on the shoaling behaviour of fish. I. Responses of two species to changes of illumination and to olfactory stimuli. J. exp. Biol. **36**, 261—280 (1959).
Stevens, P. G.: American musk. III. The scent of the common skunk. J. Amer. Chem. Soc. **67**, 407—408 (1945).
Stewart, D.: Some account of a boy born blind and deaf, collected from authentic sources of information with a few remarks and comments. Trans. roy. Soc. Edinburgh **7**, 1—78 (1815).
Stirnimann, F.: Versuche über Geschmack und Geruch am ersten Lebenstag. Jb. Kinderheilk. **146**, 211—227 (1936).
— Le goût et l'odorat du nouveau-né. Rev. franç. Pédiat. **12**, 453—485 (1936).
Stockard, C. R.: Hereditary deficiencies in the sense of smell. Science **49**, 237—239 (1919).
Stoll, M.: Les produits à odeur musquée et l'evolution de la chimie megacyclique. Chimia **2**, 217—226 (1948).
— Special problems in odor perception. Drug Cosmetic Indust. **74**, 48—49, 138—142 (1954).
— The meaning and purpose of odour. Soap, Perfumery, Cosmetics **28**, 634—637 (1955).
— Facts old and new concerning relationships between molecular structure and odour. Soc. Chem. Indust., Monograph No. 1—11. Discussion 11—12 (1957).
— De l'effet important de différences chimiques minimes sur la perception de l'odeur. Rev. Laryng. (Bordeaux) **86**, 972—981 (1965).
Stone, H.: Influence of temperature on olfactory sensitivity. J. appl. Physiol. **18**, 746—751 (1963).
— Behavioral aspects of absolute and differential olfactory sensitivity. Ann. N. Y. Acad. Sci. **166**, 527—534 (1964).
—, and J. J. Bosley: Olfactory discrimination and Weber's law. Percept. Motor Skills **20**, 657—665 (1965).
Strange, E. F.: The incense ceremony and its utensils. Trans. Proc. Jap. Soc. (London) **21**, 28—38 (1923/24).
Stratenus, E. A.: Die Vererbung der Nasenleistungen und die sich daraus ergebenden Zuchtziele. Ber. 3. kynologischer Weltkongr. 95 (1935).
Strieck, F.: Untersuchungen über den Geruchs- und Geschmackssinn der Elritze *(Phoxinus laevis, A.)*. Z. vgl. Physiol. **2**, 122—154 (1924/25).
Ströse, A.: Ein neues Verfahren zum Messen der Riechschärfe des Hundes. Jb. Jagdkunde **7**, 261—265 (1923/25).

STROHL, J.: Die Giftproduktion bei den Tieren von zoologisch-physiologischem Standpunkt zugleich ein Hinweis auf funktionelle Beziehungen zwischen Giften, Hormonen, Gerüchen. Biol. Zbl. **45**, 513—536, 578—604 (1925).

STRONG, R. M.: The sense of smell in birds. Science N. S. **24**, 943 (1908).

— On the olfactory organs and the sense of smell in birds. J. Morphol. **22**, 619—661 (1911).

STUIVER, M.: Biophysics of the sense of smell. Groningen: Dissertation 1958.

— An olfactometer with a wide range of possibilities. Acta otolaryngol. **51**, 135—142 (1960).

STÜMPKE, H.: Bau und Leben der Rhinogradentia. 3. Aufl. Stuttgart: Gustav Fischer 1962.

SUGIMOTO, B.: Kôdô. Tôkyo: Yûzankaku 1936.

SULZE: Über eine Methode zur Prüfung des Geruchssinnes des Hundes. Dtsch med. Wschr. **40**, 517 (1914).

SUMNER, D.: On testing the sense of smell. Lancet **1962 II**, 895—897.

SUMNER, J. B.: Problems in odor research from the viewpoint of the chemist. Ann. N. Y. Acad. Sci. **58**, 68—72 (1954).

SZYMANSKI, J. S.: Zur Lokalisation der osmotischen Reize. Arch. ges. Physiol. **181**, 310—315 (1920).

TAKAHASHI, T.: Arata ni kôan seru kôryoku sokutei sôchi (kôryokukei) ni tsuite. Ryôshoku Kenkyû No. 145, 531—535 (1938).

TAMMANN, G., und W. OELSEN: Zur Bestimmung der Dampfdrucke von Riechstoffen. Z. anorg. allgem. Chem. **172**, 407—413 (1928).

TANYOLAC, N. N.: The electro-odocell for odor measurement and surface effects. Bregman and Dravnieks: Surface effects in detection. Washington: Spartan Books Inc. 1965.

—, and J. R. EATON: Study of odors. J. Amer. pharm. Ass., Sci. Ed. **39**, 565 (1950).

TASMAN, A.: Meetinstrument ter beoordeeling van het reukvermogen. Chem. Weekblad **35**, 707 (1938).

TAUSSIG, S.: Verfahren zur Ausmittlung von Riechstoffkompositionen. Austrian Patent No. 94 947, 1923, November 26.

TAVERNER, P. A.: The sense of smell in birds. Auk **59**, 462—463 (1942).

TEGETMEIER, W. B.: Scentless nature of sitting birds. Bull. Brit. Ornithol. Club **15**, 88—89 (1904/05).

TEICHMANN, H.: Über die Leistung des Geruchssinnes beim Aal *(Anguilla anguilla*, L.). Z. vgl. Physiol. **42**, 206—254 (1959).

—, u. R. TEICHMANN: Untersuchungen über den Geruchsinn der Haifische. Pubb. Staz. Zool. Napoli **31**, No. 1, 76—81 (1959).

TEUBER, W.: Das Geruchsbild des Chemikers. Ätherische Öle, Riechstoffe, Parfüm., Ess. Arom. **3**, 220—223 (1953).

— Gedanken über die Ursache des Geruches und über das Zustandekommen des Geruchserlebnisses. Seifen, Öle, Fette, Wachse **79**, 373—374, 397—398 (1953).

— Ein Beitrag zur Forschung am Geruchsphänomen unter besonderer Berücksichtigung der neuen Theorie von R. M. Hainer, A. G. Emslie and Ada Jacobsen. Ätherische Öle, Riechstoffe, Parfüm. Ess. Arom. **4**, 8—11 (1954).

— Die Geruchsbegriffe der chemischen Verbindungen. Ätherische Öle, Riechstoffe, Parfüm. Ess. Arom. **4**, 57—59, 84—86, 116—118, 137—139, 161—162, 189—190 (1954).

— Dans quelle mesure peut-on établir la composition d'un parfum d'après des considérations théoriques. Indust. Parf. **11**, 98—99 (1956).

Teudt, H.: Eine Erklärung der Geruchserscheinungen. Biol. Zbl. **33**, 716—724 (1913).
— Eine neue Geruchstheorie. Wschr. Brau. **35**, 92—93, 99—100, 107—108 (1918).
— Eine neue Geruchstheorie. Dtsch. Essigindust. **23**, 1—3, 8—10, 16—17 (1919).
— Die Entstehung der Gerüche. Dtsch. Essigindust. **23**, 160—162 (1919).
— Wie entstehen die Gerüche in den Molekülen der riechenden Körper? Prometheus **30**, 201—204, 209—212 (1918/19).
— Die Entstehung der Gerüche. Wschr. Brau. **36**, 145—146 (1919).
— Ableitung der chemischen Verwandtschaft aus der Elektronentheorie. VII. Beziehungen zwischen den Valenzelektronen und den Gerüchen der chemischen Verbindungen. Z. anorg. allgem. Chem. **108**, 155—160 (1919).
— „Zur Geruchstheorie von Teudt". Prometheus **31**, 259—262 (1919/20).
— Erwiderung auf Hellers Artikel „Über die Geruchstheorie von Teudt". Biol. Zbl. **40**, 259—267 (1920).
Theimer, E. T., and J. Davies: Olfaction, musk odor and molecular properties. J. Agric. Food Chem. **15**, 6—14 (1967).
Theophrastus: Concerning odours. About 320 B. C. Greek and English translation by A. Hart. London: Heinemann 1916.
Thomas, R. E.: Selecting the modern war dog. J. Amer. Vet.-Med. Ass. **127**, 206 (1955).
Thompson, H. B.: Psychological norms in men and women. Chicago: University 1903.
Thompson, H. W.: Some comments on theories of smell. Soc. Chem. Indust., Monograph No. 1, pp. 103—115. Discussion, 115. London 1957.
Thorbjörnson, B.: Smak och lukt. Teknisk Tidskr. **66** C, 9—12. Discussion, 12—13 (1936).
Thunberg, T.: Till frågan om s. k. luktfysiologisk kompensation. Lunds Univ. Årsskr. N. S. Avd. 2, **14**, No. 5, 1—4 (1919).
Timmermans, J.: Odour and chemical constitution. Nature **174**, 235 (1954).
Toulouse, E., et N. Vaschide: Mesure de l'odorat chez l'homme et chez la femme. C. R. Soc. Biol. **51**, 381—383 (1899).
— Mesure de l'odorat chez les enfants. C. R. Soc. Biol. **51**, 487—489 (1899).
— Note sur un nouveau moyen de vérifier la loi de Weber-Fechner sur le rapport de la sensation à l'excitation et sur la vérification de cette loi par la mesure de l'odorat au moyen de solutions décimales. C. R. Soc. Biol. **51**, 640—642 (1899).
Tribe, D. E., and J. G. Gordon: Choice of diet by rats. III. The importance of the sense of smell in the choice of diets deficient in the vitamin B complex. Brit. J. Nutrit. **9**, 1—4 (1955).
Tribus, L. L.: Odors and their travel habits. Trans. Amer. Soc. Civil Engineers **85**, 378—389, Discussion 390—429 (1922).
Tryon, T.: Tryon's letters, domestick and foreign to several persons of quality. London: Conyers & Harris 1700.
Tschirch, A.: Die Riechstoffe und das Riechen. Schweiz. Apotheker-Ztg. **59**, 229—232, 241—246, 254—258, 265—272 (1921).
— Eine chemische Theorie des Riechprozesses. Schweiz. Chem.-Ztg. 105—109, 117—120, 141—144 (1922).
Tucker, D.: Electrophysiological evidence for olfactory function in birds. Nature **207**, 34—36 (1965).
Turk, A.: Olfactory sensory methods for physicochemical measurements. Microchem. J., Symp. Ser., Submicrogram Experimentation **1**, 167—180 (1961).

Tyndall, J.: Heat: a mode of motion. 7th Ed. London: Longmans 1887.
Uchida, T.: The olfactory sense of dogs and fatty acids. Proc. XIVth Internat. Congr. Zool. 1953, 292—296 (1956).
— Fatty acids and the olfactory sense of dogs. Proc. Jap. Acad. **32,** 753—758 (1956).
Ueki, S.: Problems in olfactory function. Fukuoka Acta med. **52,** 345—366 (1961).
Uexküll, J. von, u. E. G. Sarris: Das Duftfeld des Hundes (Hund und Eckstein). Z. Hundeforsch. **1,** Heft 3/4, 55—68 (1931).
Ungerer, W. G., and R. B. Stoddard: Odor value analysis. Ungerer's Bull. **3,** No. 1, 7—10 (1922).
Urbantschitsch, V.: Ueber den Einfluß einer Sinneserregung auf die übrigen Sinnesempfindungen. Arch. ges. Physiol. **42,** 155—182 (1888).
Vaglini, F.: L'Olfazione ematogena dopo polverizzazioni nasali di ialuronidasi, adrenalina, acetilcolina. Boll. Mal. Orecch. **79,** 551—567 (1961).
Valentin, A.: Ueber die Beschaffenheit der riechbaren Stoffe und die Ursachen des Riechens. Mitt. naturforsch. Ges. Bern 60—70 (1884).
Van Dam, C.: Adsorption de matières odorantes. Arch. néerl. Physiol. **1,** 666—677 (1917).
Van den Berghe, L.: Observations sur l'olfaction et sur le mécanisme des courants olfactifs chez quelques Téléostéens. Bull. Acad. roy. Belg., Classe Sci. [v], **15,** 278—305 (1929).
Van der Wielen, P.: De codificatie van kleur, smaak en reuk in de Pharmacopee. Pharm. Weekbl. Nederland **64,** 550—560, 594—599, 636—642 (1927).
Van Dishoeck, H. A. E., and N. Versteeg: On the problem of haematogenic olfaction. Acta oto-laryngol. **47,** 396—401 (1957).
Van Dorssen, S.: Het reukvermogen van eenden voor verwaaiing van den mensch. Nederl. Jager **38,** 614—615 (1933).
— Nog eens over de eenden. Nederl. Jager **39,** 26—28 (1933).
— Verwaaiing. Nederl. Jager **39,** 114—115 (1933).
Van Riper, W.: Does a hummingbird find its way to nectar through its sense of smell? Sci. American **202,** No, 2, 157—158, 160, 162, 164, 166 (1960).
Vaschide, N.: L'Expérience de Weber et l'olfaction en milieu liquide. C. R. Soc. Biol. **53,** 165—167 (1901).
— Recherches expérimentales sur l'olfaction des veillards. C. R. Acad. Sci. (Paris) **137,** 627—628 (1903).
— L'État de la sensibilité olfactive dans la veillesse. Bull. Laryngol. Otol. Rhinol. (Paris) **7,** 323—333 (1904).
— Goût. Richet: Dictionnaire de Physiologie. Vol. 7, p. 675. Paris: Alcan 1907.
—, et M. A. van Melle: Une nouvelle hypothèse sur la nature des conditions physiques de l'odorat. C. R. Acad. Sci. (Paris) **129,** 1285—1288 (1899).
Vaubel, W.: Der Geruch des Arsenwasserstoffs. Z. angew. Chem. **40,** 538 (1927).
Veress, E.: Über die Reizung des Riechorgans durch direkte Einwirkung riechender Flüssigkeiten. Arch. ges. Physiol. **95,** 368—408 (1903).
Vierling, J. S., and J. Rock: Variations in olfactory sensitivity during the menstrual cycle. J. appl. Physiol. **22,** 311—315 (1967).
Virey, J.-J.: De l'osmologie, ou histoire naturelle des odeurs, avec leur classification et des observations sur leur nature et leurs diverses modifications. Bull. Pharm. **4,** 193—228 (1812).
de Villiers, I. P.: Dogs as detectives in South Africa. Police J. **2,** 188—192 (1929).
Voge, C. I. B., and F. E. Sinclair: The Voge-Sinclair odour classification. Internat. Perfumer **2,** 70—83, 115—127 (1951/52).

Voge, C. I. B., and F. E. Sinclair: The Voge-Sinclair odour classification. III. The esters. Internat. Perfumer **3**, 42—50 (1953).

Vogt, O.: Kritisches über die Grundlagen der Hundedressur. J. Psychol. Neurol. **27**, 1—11, 96—99 (1921/22).

de Vries, H., and M. Stuiver: The absolute sensitivity of the human sense of smell. Sensory Communication. Editor: W. A. Rosenblith. New York and London: M. I. T. Press & John Wiley 1961.

Wagenaar, M.: Meetinstrument ter beoordeeling van het reukvermogen. Chem. Weekbl. **35**, 618—621 (1938).

— Eenige opmerkingen over de beoordeeling van het meetinstrument voor het reukvermogen. Chem. Weekbl. **35**, 758—759 (1938).

Wagenknecht, M.: Zucht und Dressur des Polizeihundes. Leipzig: Ehlert 1913.

Wagner, H.: Investigaciones sobre reacciones olfatorias en las aves. An. Inst. Biol. Nac. Univ. Mexico **7**, 109—117 (1936/37).

— Untersuchungen über Geruchsreaktionen bei Vögeln. J. Ornithol. **87**, 1—9 (1939).

Wahlin, A.: *See* Linnaeus.

Walker, T. J., and A. D. Hasler: Detection and determination of odors of aquatic plants by the bluntnose minnow *(Hyborhynchus notatus)* Physiol. Zool. **22**, 45—63 (1949).

Wallace, A. R.: Inherited feeling. Nature **7**, 303 (1872/73).

— Perception and instinct in the lower animals. Nature **8**, 65—66 (1873).

Wallach, O.: Bemerkungen zum chemischen Teil des Buches von Hans Henning: „Der Geruch“. Schimmels Ber. 167—184 (1922).

— Über die Ursachen für das Auftreten eines Geruchs. Dtsch. Parf.-Ztg. **9**, 9—11 (1923).

Walter, W. G.: Some experiments on the sense of smell in birds. Arch. néerl. Physiol. **27**, 1—72 (1943).

Walters, J. H.: Sight in dogs. Nature **7**, 361 (1872/73).

Warden, C. J., and L. H. Warner: The sensory capacities and intelligence of dogs, with a report on the ability of the noted dog "Fellow" to respond to verbal stimuli. Quart. Rev. Biol. **3**, 1—28 (1928).

Waterton, C.: On the faculty of scent in the vulture. Mag. Nat. Hist. **5**, 233—241 (1832).

— Remarks on Mr Audubon's "Account of the habits of the turkey buzzard (Vúltur Aúra), particularly with the view of exploding the opinion generally entertained of its extraordinary powers of smelling". Mag. Nat. Hist. **6**, 163—171 (1833).

Watson, E. R.: Smell. Biochem. J. **16**, 613—618 (1922).

Watson, W.: A letter from Mr. William Watson, F. R. S. to the Royal Society, declaring that he as well as many others have not been able to make odours pass through glass by means of electricity; and giving a particular account of Professor Bose at Wittemberg his experiment of beatification, or causing a glory to appear round a man's head by electricity. Phil. Trans. **46**, 348—356 1749/50).

— An account of Professor Winkler's experiments relating to odours passing through electrified globes and tubes, being the extract and translation from the Latin of two letters sent by that gentleman to Cromwell Mortimer, M. D. Secretary of the Royal Society. With an account of the result of some experiments made here with globes and tubes transmitted from Leipzic by Mr Winkler to the Royal Society, in order to verify the facts before-mentioned by Mr W. Watson, F. R. S. Phil. Trans. **47**, 231—241 (1751/52).

WENZEL, B. M.: Techniques in olfactometry: a critical review of the last one hundred years. Psychol. Bull. **45**, 231—247 (1948).
— Olfactometric method utilizing natural breathing in an odor-free "environment". Science **121**, 802—803 (1955).
WESSELSKI, A.: Der Sinn der Sinne. Ein Kapitel der ältesten Menschheitsgeschichte. Monogr. Archivu Orientálního, Vol. IV. Praha: Orientální Ustav 1934.
WESTER, D. H.: Eenvoudige reukmeetmethode voor gasverkenners. Subjectieve methoden voor het opsporen van strijdstoffen. Pharm. Weekbl. Nederland **74**, 476—494 (1937).
WHITE, H. H.: On testing the sense of smell. Lancet **II**, 1168 (1962).
WHITNEY, L. F.: Bloodhounds and how to train them. Revised Ed. New York: Orange Judd Pub. Co. 1955.
WILKENS, W. F., and J. D. HARTMAN: An electronic analog for the olfactory processes. Ann. N. Y. Acad. Sci. **116**, 608—612 (1964).
WILLIAMS, C. B.: Sense of smell in birds. Nature **110**, 149 (1922).
WILLIAMS, R. J.: Free and Unequal. Austin, Texas: Univ. Press 1953.
WILLIAMSON, H. C.: Pacific salmon migration: report on the tagging operations in 1925. Contrib. Canad. Biol. Fish. **3**, 267—306 (1926/27).
WINSOR, A. L., and B. KORCHIN: A device for obtaining constant olfactory stimulation. Amer. J. Psychol. **52**, 114—115 (1939).
WISBY, W. J., and A. D. HASLER: Effect of olfactory occlusion on migrating silver salmon (O. kisutch). J. Fish Res. Board Canada **11**, 472—478 (1954).
WOERDEMANN, H.: L'Influence de la température d'un gaz odorant sur la sensation olfactive. Arch. néerl. Physiol. **19**, 88—93 (1934).
— L'Influence de la teneur en vapeur d'eau d'un gaz odorant sur la sensation olfactive. Arch. néerl. Physiol. **20**, 591—595 (1935).
WOKER, G.: The relations between structure and smell in organic compounds. J. physical Chem. **10**, 455—473 (1906).
WOLFF, O. J. B.: Die Mechanik des Riechens. Berlin: C. Habel 1878. French translation in Rev. int. Sci. **2**, 422—426, 533—537, 591—595, 623—629 (1878).
WOOD, DE V.: The science of smelling. Science **20**, 215 (1892).
WREDE, W. L.: Versuche über den Artduft der Elritzen. Z. vgl. Physiol. **17**, 510—519 (1932).
WRIGHT, R. HAMILTON: Odour and molecular vibration. I. Quantum and thermodynamic considerations. J. appl. chem. **4**, 611—615 (1954).
— Odour and chemical constitution. Nature **173**, 831 (1954).
— Odour and molecular vibration. Soc. Chem. Indust., Monograph No. 1, pp. 91—102. London 1957.
— Odour and molecular vibration. Nature **190**, 1101—1102 (1961).
— Odour of optical isomers. Nature **198**, 782 (1963).
— The Science of smell. London: Allen & Unwin 1964.
— Odor and molecular vibration: the far infrared spectra of some perfume chemicals. Ann. N. Y. Acad. Sci. **116**, 552—558 (1964).
— Smells as information. New Scientist **23**, 706—709 (1964).
— Why is an odour? Nature **209**, 551—554 (1966).
— — Odour and molecular vibration. Nature **209**, 571—573 (1966).
— J. R. HUGHES, and D. E. HENDRIX: Olfactory coding. Nature **216**, 404—406 (1967).
—, —, and R. S. E. SERENIUS: Odour and molecular vibration. II. Raman spectra of substances with the nitrobenzene odour. J. appl. Chem. **4**, 615—621 (1954).

WRIGHT, R. HAMILTON, J. R. HUGHES, C. REID, and H. G. V. EVANS: Odour and molecular vibration. III. A new theory of olfactory stimulation. Chemistry and Industry 973—977 (1956).
— R. HUEY, and K. M. MICHELS: Evaluation of far infrared relations to odor by a standards similarity method. Ann. N. Y. Acad. Sci. **116**, 535—551 (1964).
WUNDER, W.: Sinnesphysiologische Untersuchungen über die Nahrungsaufnahme bei verschiedenen Knochenfischarten. Z. vgl. Physiol. **6**, 67—98 (1927).
WYBURN, G. M.: What is smell? New Scientist **5**, 524—526 (1959).
— Theory of smell. New Scientist **5**, 817 (1959).
WYLD, R. S.: The Physics and Philosophy of the Senses. London: King 1875.
"XRAYSER": The odour of sanctity. Chemist and Druggist **71**, 897 (1907).
Y, N.: Sense of orientation. Nature **7**, 483—484 (1872/73).
YOSHIDA, M.: Studies in psychometric classification of odors. IV. V. Jap. Psychol. Res. **6**, 115—124, 145—154 (1964).
— Studies in psychometric classification of odors. III. Jap. J. Psychol. **35**, 1—17 (1964).
YOUNG, C. W., D. E. PLETCHER, and N. WRIGHT: On olfaction and infrared radiation theories. Science **108**, 411—412 (1948).
ZAHN, W.: Über den Geruchssinn einiger Vögel. Z. vgl. Physiol. **19**, 785—796 (1933).
ZAVADSKY, I. V.: Circonvolution piriforme et odorat. Arch. Sci. Biol. (Moskva) **15**, 223—257 (1910).
ZELL, T.: Der Polizeihund als Gehilfe der Strafrechtsorgane. Ein Hilfsbuch für Staatsanwälte, Untersuchungsrichter, Verteidiger, Polizeibehörden usw. Berlin: Guttentag 1909.
ZIGLER, M. J., and A. H. HOLWAY: Differential sensitivity as determined by amount of olfactory substance. J. gen. Psychol. **12**, 372—382 (1935).
ZILSTORFF-PEDERSEN, K.: Further studies on blast-olfactometry. Acta oto-laryngol. **45**, 268—279 (1955).
— Die quantitative und qualitative Olfactometrie. HNO **10**, 97—102 (1962).
— Determination and variations of olfactory thresholds. Arch. Otolaryngol. **79**, 412—417 (1964).
ZWAARDEMAKER, H.: Bijdrage tot de Physiologie van den Reuk. Feestbundel van Nat. Geneesk. Onderzoek. ter Gelegenheid van den 70sten Verjaardag van F. C. Donders, pp. 179—197. Amsterdam: Van Rossen 1888.
— Die Physiologie des Geruchs. Leipzig: Engelmann 1895.
— Les sensations olfactives, leur combinaisons et compensations. Ann. psychol. **5**, 202—225 (1898).
— Geruch. Ergebn. Physiol. **1**, 2 Abt., 896—909 (1902).
— Essai d'une théorie sur l'olfactomètre. Onderz. Physiol. Lab. Utrecht Hoogesch. [v] **3**, 262—267 (1902).
— De adsorptie van muskongeur tegen vlakten van verschillend materiaal. I. II. Vers. Gewone Vergad. Wis. Natuurk., Afd. K. Akad. Wetensch. Amsterdam **16**, 31—35, 139—141 (1907).
— Over geurverwantschappen. Vers. Gewone Vergad. Wis. Natuurk. Afd. K. Akad. Wetensch. Amsterdam **16**, 183—191 (1907).
— Over geurverwantschappen, op grond van door den Heer J. Hermanides verrichte waarnemingen. Vers. Gewone Vergad. Wis. Natuurk. Afd. K. Akad. Wetensch. Amsterdam **18**, 53—60 (1909).
ZWAARDEMAKER, H.: Specifieke reukkracht en odoroscopisch ladingsverschijnsel in homologe reeksen. Vers. Gewone Vergad. Wis. Natuurk. Afd. K. Akad. Wetensch. Amsterdam **25**, 3—10 (1916).

Zwaardemaker, H.: Reukstofmengsels en hun laadvermogen door nevelelectriciteit. Vers. Gewone Vergad. Wis. Natuurk. Afd. K. Akad Wetensch. Amsterdam **25,** 512—517 (1916).
— Le phénomène de la charge des brouillards de substances odorantes. Arch. néerl. Physiol. **1,** 347—373 (1917).
— Hans Henning: „Der Geruch". Leipzig: J. A. Barth, 1916, 533 S. Arch. Ohren-, Nasen-, Kehlkopfheilk. **101,** 245—249 (1917/18).
— Over de adsorptie van riekende moleculen aan de oppervlakten van vaste lichamen. Vers. Gewone Vergad. Wis. Natuurk. Afd. K. Akad. Wetensch. Amsterdam **28,** 857—860 (1920).
— Olfactometry. Perf. Ess. Oil Rec. **12,** 308—310 (1921).
— Odeur et chimisme. Arch. néerl. Physiol. **6,** 336—354 (1921/22).
— Prüfung des Geruchsinnes und der Gerüche. Berlin und Wien: Urban & Schwarzenberg 1923. Abderhaldens Handbuch der biologischen Arbeits-Methoden. Abt. V, Teil 7.
— L'Odorat. Paris: Doin 1925.
— Odoriferous materials. Int. Crit. Tables, 1st Ed. **1,** 358—361 (1926).
— The sense of smell. Acta otolaryngol. **11,** 3—15 (1927).
— Eeen reuktheorie van het standpunt der moderne inzichten over moleculebouw. Nederl. T. Geneesk. **71,** II, 1677—1678 (1927).
—, and F. Hogewind: Het spontaan colloidaal worden van oplossingen van reukstoffen, uitermate versneld door bestraling met ultraviolet licht. Vers. Gewone Vergad. Wis. Natuurk. Afd. K. Akad. Wetensch. Amsterdam **26,** 1510—1516 (1917/18).
— — Over het teeken van het ladingsverschijnsel en den bij dit verschijnsel waargenomen invloed van lyotrope reeksen. Vers. Gewone Vergad. Wis. Natuurk. Afd. K. Akad. Wetensch. Amsterdam **27,** 69—80 (1918).
— — Le diamagnétisme des substances odorantes. Arch. néerl. Physiol. **4,** 224—232 (1919/20).
—, and K. Komuro: Contribution à la physiologie de la perception olfactive. Arch. néerl. Physiol. **9,** 261—263 (1924).
— H. R. Knoops, en M. N. van der Bijl: Het in overmaat geladen zijn van reukstofhoudende nevels, volgens proefnemingen met Heeren H. R. Knoops en M. N. Van der Bijl. Vers. Gewone Vergad. Wis. Natuurk. Afd. K. Akad. Wetensch. Amsterdam **24,** 1630—1634 (1915/16).

Biographical Notes

ALLEN, CHARLES GRANT BLAIRFINDIE (1848—1899): Writer. Born at Kingston (Ont.) Canada. Educated in America, France and England (Oxford). Author of "Physiological Aesthetics" (1877), "The Evolutionist at Large" (1881) and several works of fiction. Sometimes wrote under the pseudonyms Cecil Power and J. Arbuthnot Wilson.

ARNOLD, EDWIN (1832—1904): Poet and journalist. Graduated M. A. at Oxford in 1856, having won the Newdigate prize in 1852. Principal of the Deccan College, Bombay, 1856—1861. Editor of the "Daily Telegraph" (1873). Author of "The Light of Asia", and other poems and translations on Eastern life and thought.

AUDUBON, JOHN JAMES (1785—1851): Naturalist. Born at Les Cayes, Santo Domingo (now Haiti). Educated in U.S.A. and France. Pioneer in the drawing and painting of birds. Author of "The Birds of America" (1827—1838), "Ornithological Biography" (1831—1839) and other works. (Jean-Jacques Fougères Audubon).

AYRTON, WILLIAM EDWARD (1847—1908): Electrical engineer. Educated at University College, London. Served in the Indian Telegraph Service. Taught physics in Tokyo and London. Professor at South Kensington (1884—1908).

BONNOT, ÉTIENNE (1715—1780): Philosopher. Abbé de Mably de Condillac and titular Abbé de Mureau in the Voges. Born at Grenoble. Lifelong friend of Rousseau. Author of "Traité des Sensations" (1754), "Traité des Systèmes" (1749), "Traité des Animaux" (1755) and other works. Founder of the philosophical system known as sensationalism.

BOYLE, ROBERT (1627—1691): Natural philosopher. Fourteenth child of Richard Boyle, earl of Cork. Discoverer of "Boyle's Law" Member of the "Invisible College" which became the Royal Society of London for Improving Natural Knowledge. Author of "The Sceptical Chemist" (1661), other scientific works, and several theological treatises.

CLOQUET, HIPPOLYTE (1787—1840): Anatomist. Born in Paris where he became professor. Pioneer in rhinology. Author of "Osphrésiologie" (1821), "Traité Complet de l'Anatomie de l'Homme" (1825) and other works. His brother, Jules Germain, and his eldest son, Louis André Ernest, were also eminent anatomists.

CONDILLAC *see* BONNOT, ÉTIENNE.

DARWIN, CHARLES ROBERT (1809—1882): Naturalist. Studied at Edinburgh and Cambridge. Took part in the surveying expedition on the "Beagle" (1831 to 1836). Author of the "Origin of Species" (1859) and other related works.

DAVIES, JOHN (1569—1626): Poet. Baptized at Tibury, Wiltshire on April 16, 1569. Educated at Winchester College, Oxford. Called to the bar, 1595. Became solicitor-general in Ireland. Author of several prose works and of the philosophical poem "Nosce Teipsum".

EYQUEM, MICHEL, DE MONTAIGNE (1533—1592): Essayist. Attended the Collège de Guienne, Bordeaux and later studied law, probably at Toulouse. In his youth, freqented the French court. Author of the famous "Essays".

FISCHER, EMIL (1852—1919): Chemist. Studied at Bonn and Strasburg, Professor at Erlangen (1882), Würzburg (1885) and Berlin (1892). Foreign member of the Royal Society (1899). Nobel prizeman, 1902. Particularly famous for his work on sugars and peptides.

FOURCROY, ANTOINE FRANÇOIS, COMTE DE (1755—1809): Chemist. Studied medicine and obtained his doctor's diploma in 1780. Became lecturer in chemistry at the college of the Jardin du Roi (1784). On the day of his death (December 16) he was created a count of the French Empire.

GALTON, FRANCIS (1822—1911): Anthropologist and founder of the science of eugenics. Born near Birmingham. Graduated at Cambridge in 1844. Became a civil servant. Studied meteorology and published his "Meteorographica" in 1863. Renowned for his studies of heredity. Devised system of fingerprint identification.

GOGOL, NIKOLAI VASILIEVICH (1809—1852): Writer. Born near Mirgorod, Poltava. Left Russia in 1836 and thereafter lived chiefly in Rome. Author of "Dead Souls", "The Inspector General", "Cossack Tales" and other works.

HALLER, ALBRECHT VON (1707—1777): Anatomist and physiologist. Studied at Tübingen and Leyden. Became professor of medicine, anatomy, surgery and botany at Göttingen where he also conducted the "Göttingische gelehrte Anzeiger" to which he is said to have contributed about 12,000 papers on every branch of knowledge. Amongst his scientific works is the famous "Elementa Physiologiae Corporis Humani" (1757—1760). Also wrote philosophical romances and poems.

HEARN, LAFCADIO (1850—1904): Journalist and author. Born at Leucadia in the Ionian Islands, his mother being Greek. Acted as correspondent for American newspapers before settling in Japan where he became naturalised, changed his name to Yakumo Koizumi and married a Japanese woman. Author of "In Ghostly Japan" and many other books about that country.

HENNING, HANS (1885—1946): Psychologist. Born in Strasburg and died in Baden-Baden. Became lecturer at Frankfurt-am-Main and later (1922) professor at the Technische Hochschule in Danzig. Author of "Der Geruch" (1916 and 1924) and several other psychological works.

HEYNINX, ALBERT (1876—1955): Born at Ypres. Became M. D. of the Université Libre of Brussels and later was appointed professor in this university (1935). Author of "Essai d'Olfactique Physiologique" (1919) and of many papers on oto-rhino-laryngology. Died in Montreal.

JÄGER, GUSTAV (1832—1917): Naturalist and hygienist. Born at Bürg, Württemberg. Studied medicine at Tübingen. Became professor of zoology at Hohenheim Academy. Advocated a system of clothing associated with his name. Objected to use of vegetable fibres in clothes. Author of "Die Entdeckung der Seele" (1878), "Die Normalkleidung als Gesundheitsschutz" (1880) and other works.

LASSWITZ, KURD (1848—1910): Writer and philosopher. Born at Breslau and died at Gotha. Pioneer in the writing of science fiction. Taught in Ratibor and Gotha. Author of "Bilder aus der Zukunft" (1878) and other works.

LINNÉ, CARL VON [Linnaeus] (1707—1778): Botanist. Lectured at Uppsala. Later qualified as a medical doctor. Evolved his renowned classification system, "Systema Naturae". Visited England in 1736 and France later. Appointed professor or medicine at Uppsala in 1741 but changed to the chair of botany in the following year. Wrote many scientific books and other works.

MONTAIGNE, MICHEL DE *see* EYQUEM, MICHEL.

MYERS, CHARLES SAMUEL (1873—1946): Psychologist. Born in London. Editor of the "British Journal of Psychology" 1911—1924 and a founder of the National Institute of Industrial Psychology. Became director of the psychological laboratory of the University of Cambridge. Elected F.R.S. in 1915. For a time, professor at King's College, London. Author of a "Textbook of Experimental Psychology" (1909) and other works.

NICOLAI, GEORG FRIEDRICH (1874—1960): Physiologist, psychologist and sociologist. Born in Berlin where he became professor in 1919. Apparently because of his progressive views, his name was deleted from the list of teachers in the Friedrich-Wilhelm University in 1932. Emigrated to South America where he occupied several university positions. Author of the remarkable work "Die Biologie des Krieges" (1919) and other books, many in the spanish language.

NORTON, THOMAS (flourished about 1477): Alchemist. A native of Bristol. Seems to have been elected M. P. for his native place in 1436 and to have become a member of the Privy Council of Edward IV. Author of the "Ordinal of Alchimy" and other similar writings.

PIESSE, G. W. SEPTIMUS (1820—1882): Perfumer. Seventh child of C. A. J. Piesse, chief clerk in the War Office. Studied chemistry under Graham at University College, London. Became a partner of Wilhelm Lubin of 2, New Bond Street. Author of "The Art of Perfumery".

RÉGISMANSET, CHARLES (1877—1945): Writer. Born and died in Paris. Became a high official in the Ministère des Colonies. Author of "Philosophie des Parfums" (1907) of various romances, of essays and of books on colonial questions.

ROMANES, GEORGE JOHN (1848—1894): Biologist. Born at Kingston (Ont.), Canada. Educated at Cambridge. Elected F.R.S. in 1879. Friend of Charles Darwin. Author of "Animal Intelligence" (1881), "Mental Evolution in Animals" (1883), "Mental Evolution in Man" (1888) and other works including poems. From 1888 to 1891 he was Fullerian professor of physiology at the Royal Institution, London.

SAVELIEV, NIKOLAI ALEKSANDROVICH (1860—1943): Physiologist. Graduated in medicine at Moscow University in 1885 and became M.D. in 1892. In this year also, he took up an appointment in the University of Warsaw. Studied in Germany under Virchow and in France under Metchnikov at the Institut Pasteur. Appointed to the chair of special pathology and internal disease in the University of Yuriev (Dorpat) 1898.

THEOPHRASTUS (originally TYRTAMUS) (*c.* 372 B.C.): Philosopher. Born at Ereseus in Lesbos. Became a member of the Platonic circle in Athens. Disciple and successor of Aristotle. Author of "On the History of Plants" and other works.

TYNDALL, JOHN (1820—1893): Physicist. Born in Ireland. Professor at the Royal Institution. Colleague of and successor to Faraday. Author of "Heat Considered as a Mode of Motion" and other scientific works.

WATERTON, CHARLES (1782—1865): Naturalist and traveller. Born near Wakefield. Made four tours of South America and pioneered experiments with curare. Author of "Wanderings in South America".

ZWAARDEMAKER, HENDRIK (1857—1930): Physiologist. Professor at Utrecht. Author of "Die Physiologie des Geruchs" (1895) "L'Odorat" (1925) and many papers on olfaction.

Author Index

Subject Index

MIX
Papier aus verantwortungsvollen Quellen
Paper from responsible sources
FSC® C105338

If you have any concerns about our products,
you can contact us on
ProductSafety@springernature.com

In case Publisher is established outside the EU,
the EU authorized representative is:
Springer Nature Customer Service Center GmbH
Europaplatz 3, 69115 Heidelberg, Germany

Printed by Libri Plureos GmbH
in Hamburg, Germany